Nucleic Acid Structure and Recognition

Nucleic Acid Structure and Recognition

Stephen Neidle

The Institute of Cancer Research,
Chester Beatty Laboratories,
London SW3 6JB

UNIVERSITY PRESS

*This book has been printed digitally and produced in a standard specification
in order to ensure its continuing availability*

OXFORD
UNIVERSITY PRESS

Great Clarendon Street, Oxford OX2 6DP

Oxford University Press is a department of the University of Oxford.
It furthers the University's objective of excellence in research, scholarship,
and education by publishing worldwide in

Oxford New York

Auckland Cape Town Dar es Salaam Hong Kong Karachi
Kuala Lumpur Madrid Melbourne Mexico City Nairobi
New Delhi Shanghai Taipei Toronto
With offices in
Argentina Austria Brazil Chile Czech Republic France Greece
Guatemala Hungary Italy Japan South Korea Poland Portugal
Singapore Switzerland Thailand Turkey Ukraine Vietnam

Oxford is a registered trade mark of Oxford University Press
in the UK and in certain other countries

Published in the United States
by Inc., New York

ISBN 0-19-850635-X

Antony Rowe Ltd., Eastbourne

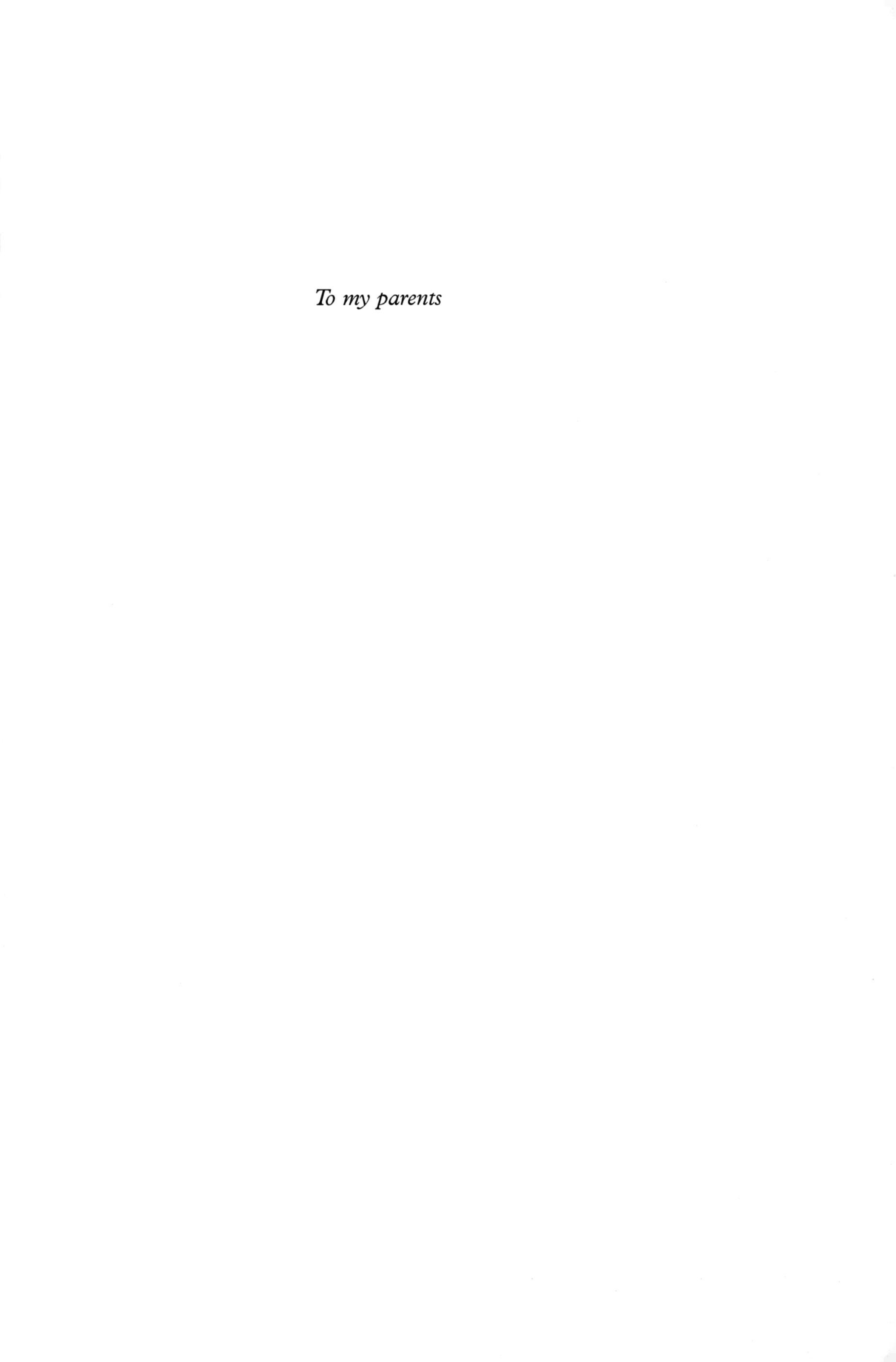

To my parents

Preface

I wrote the predecessor to this book in 1993, at a time when knowledge of nucleic acid structure was about to rapidly expand as a consequence of major technological advances in structural techniques. The Nucleic Acid Database had just been established, and RNA structural studies were still in their infancy. Seven years on, we now have many more structures, of DNAs and of large RNAs, as well as of their complexes with proteins and drugs. Of especial importance are the recent monumental achievements of ribosome and RNA polymerase structural studies. As a consequence we now have a great deal of understanding at the atomic level of many of the cellular processes involving nucleic acids, ranging from transcription and translation to DNA repair.

This book aims to provide an introduction to the underlying structural principles governing nucleic acids, and to many of the structures themselves, and thus to give a firm foundation for subsequent studies of the structural biology and chemistry of nucleic acids. Its intended audience is at final-year undergraduate or beginning graduate student level, although it is hoped that it will also be of use to active researchers. I have not attempted a comprehensive survey of all nucleic acid-containing structures. Instead, the book concentrates on those structures that illustrate a particular feature of current interest or generality, especially in the context of their relevance to biological or pharmacological issues. Such a choice of topics is inevitably somewhat subjective, and I apologize in advance to those whose favourite structure has been ignored. Inevitably, many of these choices reflect my own interests and prejudices. The book emphasizes structures determined by X-ray crystallography, since this methodology still continues to dominate the field in terms of size of molecule whose structure can be determined, as well as still providing the majority of high-resolution structures. The short introduction to this and other techniques is designed to provide the non-specialist with sufficient understanding to read the primary literature, and most importantly, to be able to begin to judge the scope and quality of structural studies.

Any book on molecular structure suffers from the disadvantage of not being able to adequately convey the three-dimensionality of structures. Rather than attempting to persuade the publishers to invest in extensive full-colour illustrations (probably a forlorn task), a dedicated internet site has been established which enables the structures to be examined interactively, and in a variety of display modes. I am very grateful to Oxford University Press for agreeing to the concept of an entirely free site, and then to putting their very considerable expertise into the project.

I am also grateful to friends and colleagues for their help, especially Struther Arnott and Helen Berman and the members of my laboratory, particularly Gary Parkinson and Martin Read for many hours of debate and discussion on the science (and sociology) of nucleic acid structural studies. I thank Cancer Research UK, which has supported work on nucleic acids in my laboratory over many years. Last but not least, I thank my wife Andrea. Writing this book would not have been possible without her constant support and encouragement.

Stephen Neidle
The Institute of Cancer Research, London
January 2002

Contents

How to use this book and its web site

Molecular structures are inherently three-dimensional. So the printed pictures in this book can only convey a limited amount of information on structures. The web site that is associated with the book will enable you to examine a wide range of representative structures in much greater detail, and will help you to better understand many of the concepts of nucleic acid structure and recognition. You will find tables listing the structures available on the web site, in the relevant chapters of the book. The original papers and the individual investigators are not cited in the tables, but you will find this information in the Nucleic Acid Database (NDB) files themselves.

The web site will be updated with new structures from time to time, but you are also encouraged to regularly browse the two primary sites for macromolecular structures, the NDB and the Protein DataBank. On entering the site you will be given a choice of several categories for structures, then a list of NDB identification codes corresponding to the codes given in the book. Clicking on a code will take you straight to a structure. We have used the publicly available CHIME plug-in, and we are very grateful to its developers for making it freely available to the scientific community.

Use of the mouse and left/right-hand buttons will enable you to move the structure around, and to change the default stick representation to one of several options. Happy viewing!

http://www.oup.com/na-structure

1

Methods for studying nucleic acid structure

1.1 Introduction

Our knowledge of DNA and RNA three-dimensional structures has advanced immeasurably since the elucidation of the first such structure, that of the DNA double helix, in 1953 by Watson and Crick in conjunction with the X-ray fibre diffraction data of Franklin and Wilkins. Fibre diffraction methods subsequently enabled the morphologies of a whole range of nucleic acid double-helical types to become established. More recently, the relationships between DNA primary sequence and fine details of molecular structure have become increasingly understood, in large part from single-crystal and nuclear magnetic resonance (NMR) structural studies on defined-sequence oligonucleotides. DNA structure itself continues to surprise with its ability to exist in a wide variety of forms, such as left-handed and multiple-stranded helices. The study of RNA structure has a more recent history, which reveals that RNA can fold in a variety of complex ways as well as occur in a double-helical form. There is also now a very large body of experimental information on the structures of protein–DNA and drug–DNA complexes.

The discovery of the double helix, as Watson and Crick realized, immediately provided fundamentally new insights into the nature of genetic events. We now have extensive knowledge of both the detail and the variety of DNA and RNA structures themselves, together with the manner in which they are recognized by regulatory, repair, and other proteins, as well as by small molecules. All this is giving us altogether more profound levels of understanding of the processes of gene regulation, transcription and translation, mutation/carcinogenesis, and drug action at the atomic and molecular levels. We are also now beginning to see this knowledge being exploited in a number of ways for therapeutic ends. For example, in conjunction with the elucidation of sequences for human and other genomes, the artificial and highly specific suppression of particular disease-associated genes with synthetic sequence-specific small-molecule ligands is becoming an attainable goal (see Chapter 5).

These advances in nucleic acid structural studies have been largely due to the increased power and sophistication of the experimental approaches of X-ray crystallography, which have provided most of the highly detailed structural information to date. The dominance of the crystallographic approach still continues, and is reflected in the emphasis of this book. NMR spectroscopy, molecular modelling/simulation and chemical/biochemical probe techniques also play important roles in providing complementary information on nucleic acid dynamics and flexibility that can approach near-atomic resolution in at least some of its detail. Traditional spectroscopic-based biophysical methods can provide important complementary information, mostly at the macroscopic level. More recently developed techniques such as surface plasmon resonance spectroscopy and atomic force microscopy are extending their power so that the gap is diminishing between macroscopic data on nucleic acids which these methods provide, and that at the atomic level from X-ray crystallography and NMR. Underlying all of this progress have been significant technical advances. These have been:

(1) in the development of routine chemical methods for oligonucleotide synthesis and purification at the milligram level for both DNA and more demanding RNA sequences;
(2) in the advent of efficient (and increasingly routine) cloning and expression systems for RNA and DNA-binding proteins, and for native RNA molecules.

This chapter provides a brief introduction to these various structural methods, emphasizing their scope as well as their limitations for nucleic acid structural studies.

1.2 X-ray diffraction methods for structural analysis

1.2.1 Overview

X-rays typically have a wavelength of the same dimensions as interatomic bonds in molecules (about 1.5 Å). Scattering (or diffraction) of X-rays by molecules in ordered matter is the result of interactions between the radiation and the electron distribution of each component atom. Typical diffraction patterns from DNA, in the form of fibres or single crystals, are shown in Figs 1.1 and 1.2. Reconstruction of the internal molecular arrangement by analysis of the scattered X-rays, analogous to a lens focusing scattered light from a microscope sample, thus provides a picture of the electron density distribution in the molecule. This reconstruction is not generally straightforward, due to the loss of phase information from the individually reflected X-rays during the diffraction process. The phase problem needs to be solved (see Section 1.2.2 for a brief description of various methods of doing this) in order for the electron density to be calculated in three dimensions (as a Fourier series), which is commonly termed a Fourier map. The approximate equivalence of the wavelength of X-rays and bond distance, of ~1–1.5 Å, means that in principle, the electron density of individual atoms in a molecule can be resolved, provided the pattern of diffracted X-rays can be reconstituted into a real-space image.

The degree of electron-density detail that can actually be seen is dependent on the resolution of the diffraction pattern observed. Resolution may be defined in terms of the shortest separation between objects (i.e. atoms or groups of atoms in a molecule) that can be observed in the electron density reconstituted from the diffraction pattern. The resolution limit (r) is governed by the maximum diffraction angle (θ) recorded for the diffraction data and the wavelength (λ) of the X-rays: r is defined as $\lambda/2 \sin \theta$. At a resolution of 2.5 Å, individual atoms in a structure cannot be resolved in an electron-density map although the shape and orientation of ring (e.g. base pairs) systems can be readily distinguished. They appear as elongated regions of electron density, with substituents

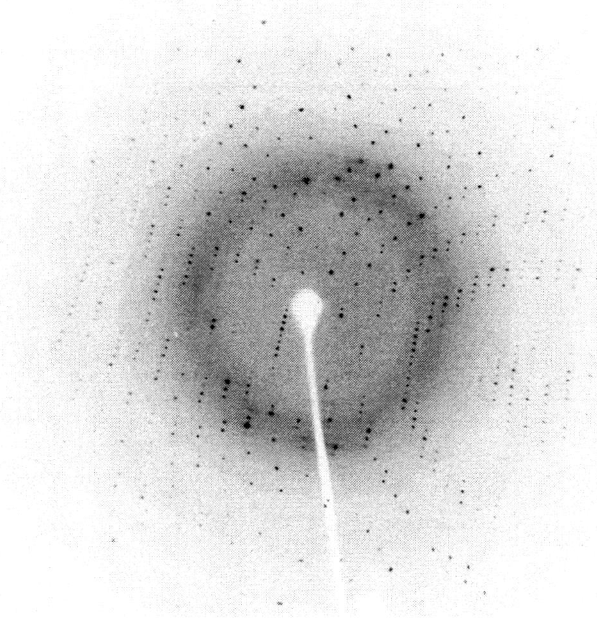

Fig. 1.1 X-ray diffraction pattern from a single crystal of a drug–oligonucleotide complex, taken with an image plate and a conventional laboratory X-ray source. The resolution limit for this pattern is about 1.7 Å. I am grateful to Michael Lee (Chester Beatty Laboratories) for providing this picture.

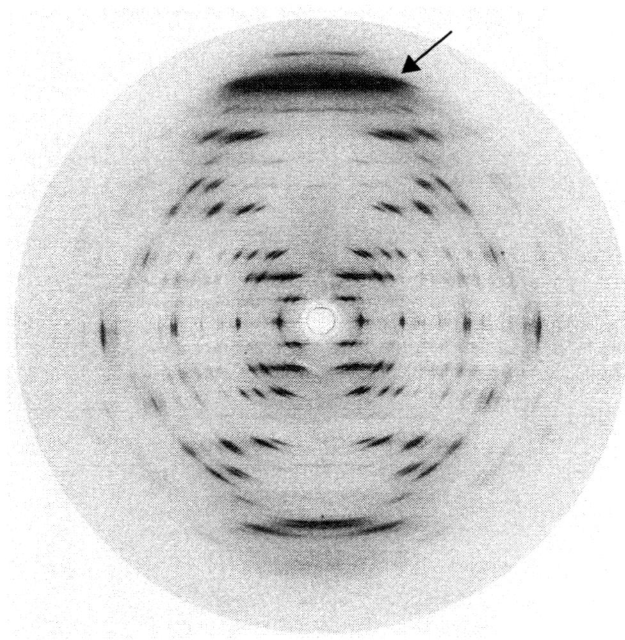

Fig. 1.2 X-ray fibre diffraction pattern from a sample of calf thymus DNA, showing a characteristic B-form pattern. The 3.4 Å layer line is arrowed. Struther Arnott is thanked for providing this picture.

being apparent as 'outgrowths' from the main density. At 1.5 Å, individual atoms are generally just about observable in a map, although only at about 1.0–1.2 Å are all atoms fully resolved and separated from each other (Fig. 1.3). There has been a marked increase in high-resolution studies (up to 0.8 Å) in recent years due to the increasing use and availability worldwide of high-flux synchrotron sources of X-rays for structural biology studies. These have intensities of X-ray beams greater by several orders of magnitude than conventional laboratory X-ray sources. Synchrotron facilities have also enabled much smaller crystals than hitherto to be successfully analysed. Most importantly, the ability to tune X-rays to differing wavelengths has provided the means whereby powerful methods of structure analysis can be employed (see below). It is now almost universal practice to collect diffraction data from macromolecules at liquid nitrogen temperatures using the technique of 'flash-freezing', which tends to minimize crystal decay in the X-ray beam, and can improve the diffracting power of a particular crystal.

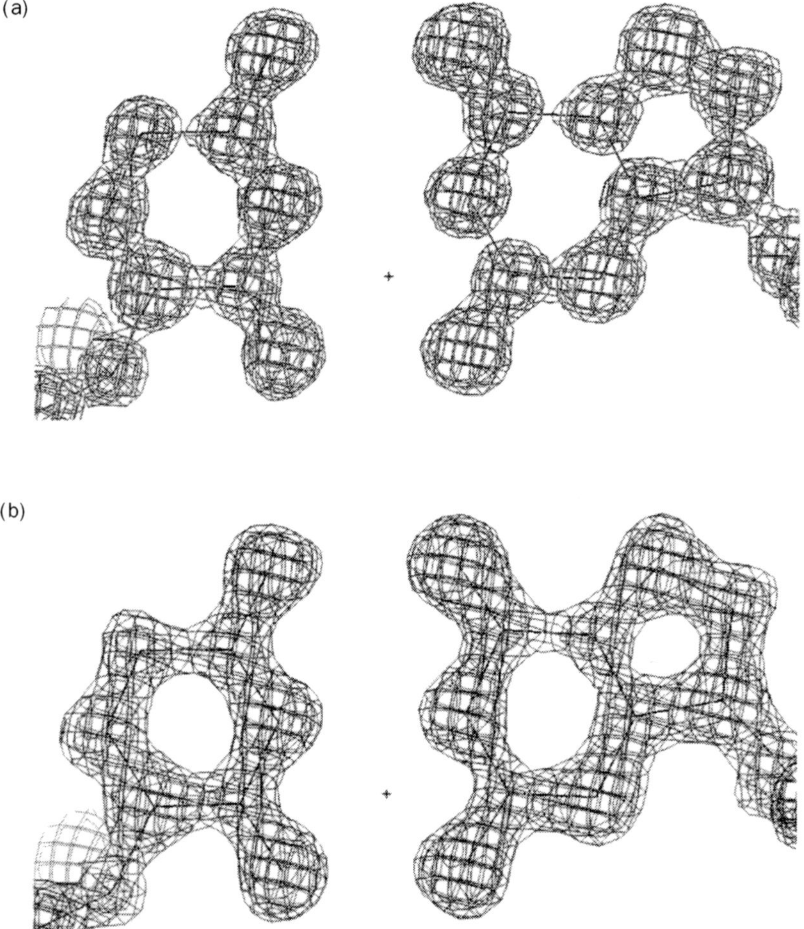

Fig. 1.3 Calculated electron density in the plane of a C•G base pair, calculated at differing resolutions, showing the amount of atomic detail visible at particular resolutions. (a) 0.9 Å, (b) 1.5 Å, (c) 2.0 Å, (d) 2.5 Å.

(c)

(d)

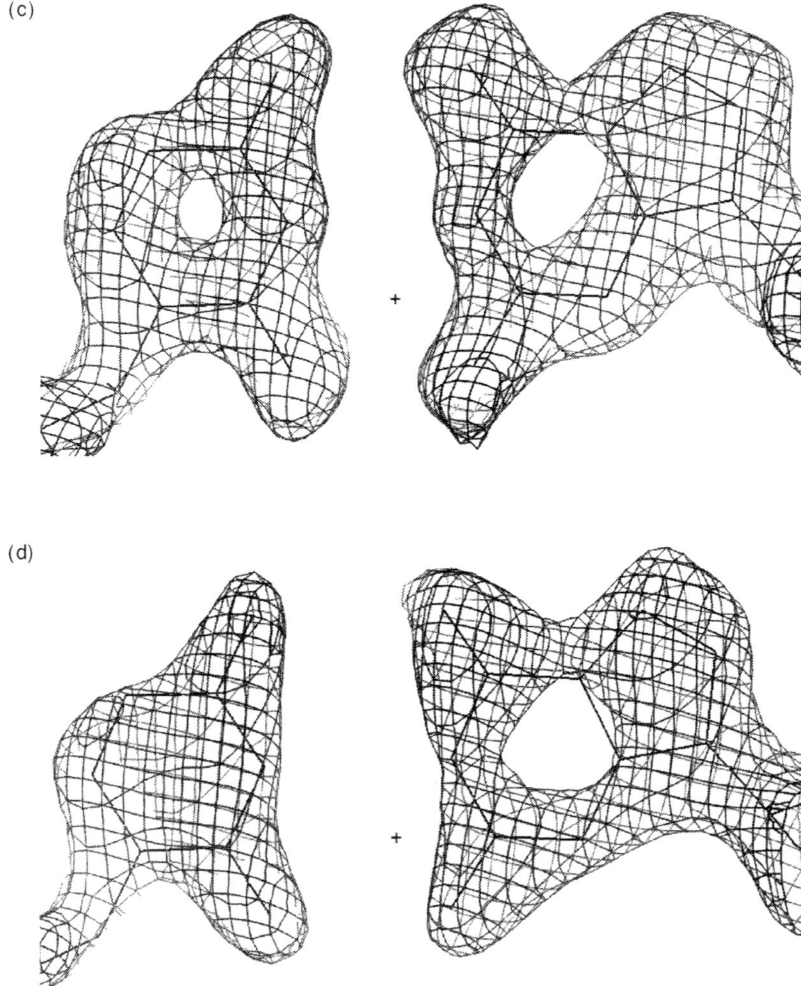

Fig. 1.3 (Continued)

The number of individual diffraction maxima observed from a crystalline or semi-crystalline sample depends directly on the resolution of the pattern. The number of unique reflections derived from these measurements also depends on the symmetry of the crystal. An ultra-high-resolution (0.74 Å) structure of a typical DNA 10-mer oligonucleotide crystal would give approximately 29000 individual unique maxima (reflections) in a monoclinic space group. By contrast, a 12-mer oligonucleotide crystal in the well-studied orthorhombic space group $P2_12_12_1$, would give only some 3000 unique reflections at 2.2 Å resolution, which until recently, was a common limit for oligonucleotide crystals. The measured intensity of an individual reflection is proportional to its structure amplitude, or observed structure factor, which when combined with phase information, results in the calculated structure factor. X-ray structures are always optimized (refined) against this observed structure amplitude data, usually by a least-squares method (see below).

The accuracy and reliability of the resulting structure depends in part on the quantity and resolution of the diffraction data, as well as the quality of its measurement. Of key importance is the actual correctness of the structural model itself, both in gross outline (which is defined by the low- to medium-resolution data), and in the detailed aspects of the structure (defined by the high-resolution data). The standard ways of assessing these factors for a given structural model are:

(1) to calculate the crystallographic R factor, defined as: $R = \Sigma||F_{\mathrm{o}}| - |F_{\mathrm{c}}||/\Sigma|F_{\mathrm{o}}|$, summed over all observed reflections, where F_{o} and F_{c} are the observed and calculated structure factors. R, which is also termed the reliability index, is expressed as a percentage, or sometimes as a decimal;

(2) to calculate the so-called free R factor (R_{free}) for a small (typically 5 per cent) subset of reflections, often chosen randomly. This set is not used in the refinement, and so the value of R_{free} is unbiased by the course of the refinement and any errors introduced in it.

The R value for a correct structural model can range from <10 to >20 per cent; in general the lower the value the more reliable is the model. Values for R_{free} are usually a few per cent higher, but are very sensitive to even small changes and errors in the model. It is often used to judge the completion of a structure analysis in terms of the behaviour of water molecules located in electron-density maps and added to the model during successive rounds of refinement. At a certain point, adding more water molecules may well reduce the value of R simply because the number of variables is increased in the least-squares refinement; however if the value of R_{free} increases, then these 'water molecules' are not real. It is also common practice to calculate Fourier maps with parts of a structure omitted to verify that these parts re-appear in the map at the stereochemically correct positions.

1.2.2 **Fibre diffraction methods**

Historically, helical DNA and RNA structures were first analysed by fibre diffraction techniques. Polymeric nucleic acids directly extracted from cell nuclei, have not been crystallized as single crystals capable of three-dimensional structure analysis. Initial diffraction patterns of poor quality were obtained from such DNA by Astbury in 1937/38, but significant progress was not made until the early 1950s, by Wilkins, Franklin and their associates. They made DNA into oriented fibres, when the act of 'pulling' such a fibre orients the nucleic acid helix along the direction of the fibre. These fibres can have exactly repetitious helical dimensions even though the underlying naturally occurring genomic nucleic acid sequences in them are not simple repeats, and thus the sequence information in the nucleic acid molecules is lost. An important exception occurs with the use of synthetic polynucleotides of known, simply repeating sequences such as poly(dA–dT)•poly(dA–dT).

Natural and synthetic polynucleotides can form fibres with varying degrees of internal order, having one- or two-dimensional paracrystalline arrays in the fibre, with the latter usually having the greater order because of their non-random sequences. These differing degrees of order are reflected in their X-ray diffraction patterns, with natural, double-helical DNA and RNA molecules usually having a degree of order along the helix axis but being randomly oriented with respect to each other. This gives rise to an X-ray diffraction pattern with characteristic spots and streaks of intensity—for example, the 'helical cross' diffraction pattern, which is characteristic of B-type DNA double helices. Such patterns can be analysed to give the helical dimensions of pitch, rise and number of residues per helical turn, as well as defining the overall helical type (A, B, etc.). Even the best-ordered of paracrystalline polynucleotide fibres give at most only a few hundred individual diffraction maxima, corresponding to a maximum resolution of about 2.5 Å.

It is not in general possible to analyse this pattern of fibre diffraction intensities, determine phases and derive a molecular structure *ab initio*, since the pattern is an average from all of the nucleotide units in a helical repeat. Instead, the pattern is fitted to a model using a least-squares procedure, which enables conformational details of the averaged (mono- or di-nucleotide) repeat to be varied and optimized. The correctness and quality of the model may be assessed using the standard crystallographic R factor. Values for R of 0.15–0.25 indicate that the calculated diffraction pattern agrees well with the observed one, and that the model is physically reasonable in terms of its stereochemistry. An important question is whether the attainment of good agreement for these criteria necessarily means that the phase problem for these fibre structures has been uniquely solved. The process of analysis assumes a particular starting model and other models might in principle fit a data set at least moderately well, especially since atomic levels of resolution are not available from fibre diffraction. This question led, some years ago, to several suggestions of alternative structures to the Watson–Crick antiparallel double helix for DNA. Only on detailed examination was it found that none of these alternatives could be fitted in an acceptable manner to the observed diffraction data, as defined by the R factor and other tests. This, together with their numerous close intramolecular contacts, enabled these alternative structures to be conclusively rejected and the antiparallel double helix accepted as the sole model that can acceptably fit the observed B-DNA diffraction data.

It is striking, in spite of the limitations of fibre diffraction methods, that their characterizations of idealized DNA and RNA double-helical geometry have been found to be closely in accord with more recent single-crystal analyses of short helical segments, even at much higher resolutions. Fibre diffraction analysis has also the considerable advantage of being able to readily study conformational transitions under a range of environmental conditions. The A↔B transition of duplex DNA observed with variations in relative humidity is the classic example of this technique.

1.2.3 Single-crystal methods

By contrast, single-crystal X-ray crystallographic analyses are able to determine the complete three-dimensional molecular structures of biological macromolecules without necessary recourse to any preconceived model, provided the molecules are discrete and not the effectively infinite polymers of nucleic acid fibres. Single crystals can be thus defined as ordered arrays of discrete and identical molecules in three dimensions.

Crystallization of many DNA and RNA oligonucleotides has historically been challenging, being sometimes more dependent on chance than systematic scientific study. This has changed radically with the advent of manual (and increasingly) automated methods to rapidly and systematically screen a wide set of crystallizing conditions. A number of commercial kits are now available with pre-prepared solutions of a large range of concentrations and types of counter-ion, buffer and precipitating agents, so that a large number of crystallizing trials can be set up with minimal effort. This approach is also useful for finding alternative crystal forms if initial trials produce crystals with poor diffraction or exceptionally large unit cells. The increasing use of robotic crystallization methods is enabling rapid, large-scale screening of crystallization conditions to be undertaken, which is especially useful when dealing with 'difficult' molecules such as large RNAs or protein–nucleic acid complexes.

The range of resolution reported for single-crystal studies of oligonucleotides spans from 0.7 to 3.0 Å (Fig. 1.3). Thus, the highest-resolution oligonucleotide structures have true atomic resolution and accordingly are of corresponding accuracy ($\leqslant$0.02 Å for distances and $\leqslant$0.2° for angles) in respect of derived geometric parameters. A typical 2.5 Å resolution structure analysis, by contrast, would have distances reliable to about $\pm$0.3 Å and angles to about $\pm$5°. However, it is necessary to use constraints to standard bond geometries during the crystallographic refinement process of non-atomic resolution crystal structures. This means that it is not only non-bonded and intermolecular distances but also conformational and base morphological features that have to be interpreted with care, and likely errors and uncertainties taken into account. Hydrogen atoms are only directly observed in electron-density maps from the highest-resolution oligonucleotide analyses, and so hydrogen-bonding schemes (especially those involving water molecules) are normally only inferred.

X-ray diffraction patterns from oligonucleotide crystals can be analysed, and their underlying molecular structures solved *ab initio* by the standard heavy-atom multiple and single isomorphous replacement (MIR and SIR) phasing methods of macromolecular crystallography. These do not presume any particular structural model and hence do not bias the resulting structure, for example, to have all Watson–Crick base pairing in a double-stranded oligonucleotide. However, a number of such heavy-atom derivatives are required for satisfactory MIR phasing, which are not always readily obtained, especially for helical nucleic acids. In favourable cases, it is possible to solve a structure with a single derivative by means of a combination of phasing from isomorphous replacement and anomalous scattering at a single wavelength. The advent of tunable-wavelength X-ray facilities at high-flux synchrotron facilities has enabled the technique of phasing by multiwavelength anomalous diffraction (MAD) to be used. This uses a single appropriate heavy atom, which has the ability to absorb X-rays to differing extents at different wavelengths; phases and hence electron density maps can be directly calculated from such data. These maps, when obtained at high resolution, are sometimes of remarkably high quality, revealing complete structures at the outset. It is fortunate for nucleic acid crystallography that bromine atoms, which can be readily chemically attached to uracil bases, provide excellent anomalous diffraction signals. An increasing number of structures are being solved by this powerful method, with the principal limitation being the availability of sufficient, tunable synchrotron beam lines. Alternatively, synthesis of an analogous sequence containing a single nucleotide with a thio-containing backbone can be used to bind a mercury heavy-atom derivative.

Alternatively, it is possible to take account of the fact that many nucleic acid structures crystallize in an isomorphous arrangement to ones previously determined (e.g. by heavy-atom phasing) or are presumed to contain a particular structural motif such as a double helix. These structures can often be solved by molecular replacement or 'search' methods, which assume at least part of the structure and attempt to locate it in the crystallographic unit cell. Problems have occasionally arisen with this approach, when a helix has been correctly oriented within the unit cell but its position is incorrectly indicated, being systematically related to the correct one, for example, by a simple translation of a base pair. Search methods become increasingly challenging with a decreasing fraction of known geometry in a structure, and heavy-atom methods then become advisable. They are also difficult when the correct geometry of the search fragment is not precisely known, and then the correct rotational and translational solution becomes unclear. New protein–nucleic acid crystal structures are usually solved by heavy-atom MIR and MAD methods, as are an increasing number of new types of oligonucleotide crystal structures. It is fortunate that the key crystal structures of a B- and a Z-DNA oligonucleotide have been solved *ab initio* by heavy-atom methods (see Chapter 3), thereby ensuring a firm and unambiguous basis for subsequent molecular replacement analyses of a large number of DNA oligonucleotide structures.

The increasing possibility of obtaining true atomic-resolution ultra-high quality synchrotron diffraction data on a few oligonucleotides, whose crystals are exceptionally well-ordered and diffract to better than 1.0 Å, provides the opportunity for phasing methods that do not rely on any heavy atoms being required. Several pioneering studies have shown that 'direct' methods, which employ mathematical relationships between phases, may be used in favourable cases to compute phases and then electron-density maps from native structures. The anomalous signal from phosphorus can also be used as a 'heavy atom' to obtain phases, again only in the most favourable of circumstances.

Macromolecular crystal structures are normally optimized with respect to the diffraction data by non-linear least-squares fitting procedures, which formally minimize the differences between observed and calculated models for the structure factors. This is the process of crystallographic refinement. When the diffraction data does not extend to atomic resolution, it is necessary to incorporate information from established stereo-chemical and structural features (such as bond lengths and angles, planar geometry of the DNA bases, preferred torsion angles). These are used to set up intramolecular constraints and restraints between them and so improve the initial models. The most widely used programs for macromolecular refinement, X-PLOR and CNS, use empirical energy terms as part of the minimized function to ensure optimal intra- and inter-molecular geometry. The technique of simulated annealing has been adopted from molecular dynamics as an effective way of refining structures when large scale (>1 Å) atomic movements are required, since conventional least-squares methods are inherently incapable of effecting such large changes.

Oligonucleotide and oligonucleotide–protein crystals are heavily hydrated, with often over 50 per cent solvent content. It is typical in medium-resolution structures for only a small fraction of these water molecules to be located in electron-density maps, largely because their high mobility smears their electron density to below the signal-to-noise level of these maps. The majority of water molecules reported in these structures are unsurprisingly the least mobile ones, which are directly hydrogen-bonded to the structure—these are the 'first-shell' water molecules (see Chapters 3 and 5 for detailed discussions of water arrangements in nucleic acid structures). The ways in which molecules pack in the crystal are sometimes of importance when examining structural features, since considerations of efficient packing can readily force parts of molecules to interact with one another by hydrogen bonding and van der Waals interactions, and consequently possibly modify some features of an otherwise flexible conformation.

The quality and reliability of an oligonucleotide crystal structure are not straightforward to assess, especially for a non-crystallographer. Yet, judgements on these factors are critical when undertaking and using structural comparisons and analyses. The important crystallographic parameters of quality (R, R_{free}) have been outlined above. Of at least equal significance are the derived stereochemical features—examination of these is a reliable guide to quality.

Particular features to examine in a structure include:

(1) close non-bonded intra- and intermolecular contacts that are less than the sum of the van der Waals radii of the atoms involved;
(2) the distribution of values for torsion angles around single bonds. Eclipsed (~0°) values are indicative of problems in refinement;
(3) hydrogen bonds with distances appreciably outside the accepted ranges of ~2.7–3.2 Å;
(4) estimates of error in atomic positions;
(5) the quality of the electron density for individual groups and atoms;
(6) values for atomic temperature factors, especially for water molecules.

These purely structural factors are equally applicable when examining structural models derived from NMR analyses.

1.3 NMR methods for studying nucleic acid structure and dynamics

The underlying principle of nuclear magnetic resonance is the detection, in a magnetic field, of those atomic nuclei in a molecule which have nuclear spin. Protons are abundant in nucleic acids and oligonucleotides, and fortunately have readily detectable spin signals. These signals, termed chemical shifts, are dependent on the shielding effect of neighbouring protons, and thus can be used to determine the chemical environment of a proton once they can be unequivocally assigned as arising from particular atoms. Examples of highly characteristic, conformation-dependent chemical shifts are those arising from the protons on a deoxyribose sugar, which vary according to the pucker of the sugar (see Chapter 3). Other nuclei are less sensitive than protons, and have low natural abundance but the ever increasing field strengths of available magnets used in NMR studies are making ^{13}C and ^{15}N-enriched oligonucleotides amenable to detailed studies. NMR studies of oligonucleotides have been extensively used to examine interactions, by monitoring characteristic changes in particular chemical shifts.

The magnetic interactions between a pair of protons gives rise to an NMR spin–spin coupling constant which is directly related to the dihedral angle between them, by the Karplus relationship. Hence, measurement of coupling constants provides direct and reliable information on sugar puckers and on part of the backbone conformation (notably the glycosidic angle between sugar and base) in a nucleotide or oligonucleotide. Recent use of ^{13}C and ^{13}N labelled oligonucleotides has enabled coupling constants to be determined for otherwise inaccessible backbone torsion angles.

NMR methods enable structures to be determined in solution, largely by means of measurements of proton–proton coupling constants and through-space nuclear Overhauser effect (nOe) derived distances using 2D NMR methods. Solution-phase studies have the obvious advantage that molecules do not have to be crystallized, which is often the major (and highly frustrating!) limitation to the analysis of a macromolecule by X-ray crystallography. There is also the apparent advantage that a structure determined in solution is more relevant to physiological processes than an X-ray crystallographic study in the solid state. However, the two techniques should not be considered as alternatives. Rather they are complementary, providing distinct information. For example, NMR results emphasize the flexible nature of DNA molecules and the fact that individual groups such as sugars are dynamically in motion. It is notable that parallel observations of sequence-dependent effects in a number of oligonucleotide sequences have been reported from both crystallographic and NMR studies, although differences in detail are sometimes apparent.

There are a number of limitations to the accuracy and reliability of NMR methods as applied to nucleic acids. By contrast with crystallography, there is a limitation on the size of the problem that can be analysed in detail, due to the increase in the number of signals with molecular weight, and consequent overlap of chemical shifts. The nOe is significant only for protons, and so phosphate geometry is not directly defined by it. Hence, nucleotide backbone conformations are, in principle, incompletely defined by standard ^{1}H NMR methods. Since a nOe intensity is proportional to the inverse sixth power of a proton–proton distance, it is a short-range effect, which is only significant at distances less than 5–6 Å. Longer distances important in DNA structure, such as groove width, may not be derived directly. Most reliability from NMR experiments can be placed on particular features of DNA structure, especially base pairing, sugar pucker and location of ligand binding sites. Detailed aspects of, in particular, sequence-dependent structure have been, until recently, a matter of considerable controversy. This is, in large part, because of the relatively small number of nOe data available compared to the several thousands of X-ray intensities from a typical medium resolution crystallographic analysis. The consequent under-determination of NMR-derived DNA structures means that their effective 'resolution' is probably ~3–4 Å, with those parts of a structure providing the most nOe data being the most reliable. This situation is distinct from that for small ($<$25 kD) globular proteins, where the richness of nOe data arising from compact, closely-packed amino acid residues, provides for highly reliable and detailed NMR structure determination more equivalent to that from high-resolution crystallographic analyses.

A standard approach in NMR structure determination is to use restrained molecular dynamics methods and an assumed rough starting model. The most commonly used programs are X-PLOR and CNS, thus sharing common parameters and algorithms with crystallographic refinements. The distances established from individual nOe assignments are taken as constraints together with the NMR-derived conformational angles, to arrive at a plausible model or set of models. It is common practice to employ the nOe data in a semi-quantitative manner, with three groups of distances being used. These are long (e.g. 4–6 Å), medium (3–5 Å) and short (2–3.5 Å). The accuracy of an NMR structure can be assessed by back-calculating the nOe intensities and comparing them with the observed, in a manner analogous to that used in crystallography. However, the NMR 'R factor' can be a less reliable guide since the number of observations is so much less than in crystallography, and flexibility in part of a structure may bias R to give an apparently poor value.

1.4 Molecular modelling and simulation of nucleic acids

Crystallographic analyses provide a quasi-static view of molecular structure. The process of X-ray data acquisition from a single crystal, even at a high-flux synchrotron source, can take typically several minutes. It thus provides a time-averaged picture of molecular motions about the low-energy structure in the crystal (which is often $>$50 per cent solvent). By contrast, molecular modelling techniques enable dynamic changes in structure and conformation to be calculated and visualized in terms of their effects on molecular energetics. The theoretical methods thus provide information complementary to the experimental techniques.

It is not feasible at present to compute conformational or energetic properties for lengths of nucleic acid sequence by *ab initio* quantum mechanics. Instead, empirical force-field methods are widely used. These have been derived from experimental data that describe the energetics of a DNA or RNA molecule in terms of the sum of a number of factors:

- van der Waals non-bonded interactions
- bond length and angle distortions
- barriers to rotation about single bonds
- electrostatic contributions from full and partial electrostatic-potential derived atomic charges
- hydrogen bonding.

The most recently developed nucleic acid force fields can also incorporate the contributions of polarization effects, which can be especially important when examining the interactions of nucleic acids with drug or protein molecules.

The major empirical nucleic acid force-fields have been incorporated into algorithms that minimize the conformation of a molecule with respect to its internal energy. This is the method of molecular mechanics, which in effect optimizes local low-energy minima. Much more extensive explorations of conformations can be made by molecular dynamics (MD), which applies Newton's equations of motion to an empirical force-field, for all atoms in a molecule. This technique is computer-intensive, although with current desktop workstations (or even with high-end PCs) it is possible to undertake realistic simulations of molecular motions, with the inclusion of large numbers of calculated solvent molecules. The widespread availability of supercomputers has enabled an increasing number of simulations to be performed over thousands of picoseconds of molecular movements, with a number of studies of >10 nanoseconds (1 nanosecond = 1000 picosecond) duration. This timescale is still far from that of biochemical events, but the rate of increase in computer power makes true biological simulations an attainable goal in the future.

Molecular dynamics can enable barriers between local energy minima to be traversed, unlike molecular mechanics. It is widely used in conjunction with distance geometry data (from nuclear Overhauser effects) to derive plausible DNA/RNA and nucleic acid–protein structures from these NMR measurements. The technique of simulated annealing is often employed, in which a structure is first simulated by molecular dynamics at a high temperature, when it is able to overcome high-energy barriers between conformations. The system is then gradually 'cooled', when the most likely energy states become populated.

Solvent and counter-ions are normally, routinely incorporated into dynamics simulations, their positions having been generated by Monte Carlo algorithms. The earlier use of a distance-dependent dielectric model to compensate for the lack of solvent often led to unstable structures during MD simulation. Even with the inclusion of a simple solvent model, nucleic acid simulations are sometimes difficult to maintain stable due to the inability of the force field to adequately account for long-range electrostatic forces and solvation effects. This is seen in the breaking apart of structures after a period of simulation, which can be several hundred picoseconds. The recent introduction of Ewald and particle-mesh Ewald summation methods has led to a marked improvement in the stability of nucleic acid systems even in long time-scale simulations. Validation against both NMR and X-ray crystallographic structures is an important activity, especially with the advent of a new generation of accurate and reliable experimental structures. Simulations with explicit solvent are still computationally expensive when applied to large systems. The use of the Generalized Born solvation model enables DNA to be modelled without the need for explicit water and counter-ions molecules to be present. This treatment can adequately account for solvent continuum effects and solute–solvent electrostatic polarization.

The use of molecular mechanics and dynamics methods have greatly increased in recent years, due not only to the ready availability of high-performance computing facilities, but also to the widespread availability of packaged commercially derived computer programs with graphical front-ends for the display of results. There are a number of modelling programs in common use which have their force fields parameterized for nucleic acids and their components.

The most widely used and tested simulation programs are:

1. AMBER (Assisted Model Building and Energy Refinement), from the laboratory of P. A. Kollman, University of California, San Francisco. The second-generation 'Cornell *et al.*' force-field incorporated in recent releases of this package, have been especially well-validated for nucleic acids.
2. CHARMM (Chemistry at Harvard Macromolecular Mechanics), from the laboratory of M. Karplus, Harvard University. The forerunner of X-PLOR and CNS.
3. GROMOS (Groningen Molecular Simulation), from the laboratory of W. F. van Gunsteren and H. J. C. Berendsen, University of Groningen, The Netherlands.
4. JUMNA (Junction Minimization of Nucleic Acids) from the laboratory of Richard Lavery, Institut de Biologie Physico-Chimique, France. This program enables the torsion angles and helicoidal parameters in a structure to be varied, rather than standard x, y, z cartesian coordinates. It enables large regions of conformational space to be rapidly explored.

1.5 Chemical and enzymatic probes of structure

Enzymes such as DNase I cleave the phosphodiester bond in a DNA duplex at every nucleotide position, although the cutting efficiency is markedly dependent on sequence, and by implication, on sequence-related structural features. Cleavage may be blocked by protein or drug binding. Hence DNase I can be used to determine sites of binding along a DNA sequence as well as to assess possible effects of particular sequences on DNA structure. Chemical cleaving agents such as hydroxyl radicals, can give similar information. Since these are much smaller molecules than cleavage enzymes, their effects on DNA structure are less perturbing and sequence-dependent. Other types of chemical probe can attack specific base sites. These can be useful in defining the precise sites of protection resulting from drug or protein binding to a DNA sequence.

These methods have the important advantage over the fine-structure techniques of crystallography and NMR, of being applicable to long (up to several thousand base pair) DNA sequences, and thus of being more directly relevant to DNA in the cell. Hence, the use of chemical and enzymatic probes for DNA provides a way of obtaining at least some molecular-level data on otherwise inaccessible structural problems in DNA–protein and drug recognition.

1.6 Sources of structural data

The results of a crystal structure or fibre diffraction analysis are most useful as a set of atomic coordinates. Those from fibre diffraction are available either in the primary literature or in various review chapters and compilations (see Chapter 3). Crystallographic coordinates may be obtained from a database, provided they have been deposited in the first instance. This is no longer a problem since almost all journals now insist on deposition by publication, although authors sometimes retain the right to up to a year's delay before public release of a data set. All available oligonucleotide crystal structures are in the successor to the Brookhaven Protein Data Bank, the RCSB (Research Collaboration for Structural Biology) Protein Data Bank. Many depositions are accompanied by structure factor data. This Data Bank also contains a number of modelled structures. The Cambridge Crystallographic Database (which is primarily for small molecules) also contains coordinate data on a number of oligonucleotide structures. The Nucleic Acid Database (NDB) is a comprehensive relational database for nucleic acid crystallographic data, at Rutgers University, USA. It also provides a set of powerful tools for the comparative study of nucleic acid structural features, enabling detailed analysis of trends and features to be readily undertaken. Both the RCSB and the NDB contain all deposited protein–nucleic acid crystal structures, as well as many NMR-derived ones. Many NMR structures are also available in the BioMagResBank Repository.

These databases are accessible via the internet from their primary sites, or from a number of mirror sites worldwide. The mirror sites for the NDB are detailed at the end of this chapter.

Further reading

General

Blackburn, G. M. and Gait, M. J., (ed.) (1996). *Nucleic Acids in Chemistry and Biology*, 2nd edn, Oxford University Press, Oxford.
A useful survey of nucleic acid function, with some emphasis on structural aspects.

Bloomfield, V. A., Crothers, D. M., and Tinoco, I., Jr. (2000). *Nucleic Acids. Structures, Properties and Functions.* University Science Books, Sausalito, California.
An indispensable book, emphasizing the biophysical chemistry of nucleic acids, but covering many structural aspects as well.

Calladine, C. R. and Drew, H. R. (1997). *Understanding DNA*, 2nd edn, Academic Press, London.
An entertaining account of selected areas of DNA structure.

Lilley, D. M. J. and Dahlberg, J. E. (ed.) (1992). *Methods in Enzymology: DNA Structures*. Volumes 211, 212. Academic Press, San Diego.
A comprehensive compilation of reviews on many aspects of DNA structure, emphasizing the methodologies of the early 1990s.

Neidle, S. (ed.) (1999). *Oxford Handbook of Nucleic Acid Structure*. Oxford University Press, Oxford.
A detailed reference book on atomic-level DNA and RNA structures in the crystalline and solution states as studied by X-ray and NMR methods.

Olby, R. (1974). *The Path to the Double Helix*. Macmillan Press, London.
An authoritative and detailed account of the history of the discovery of the double helical structure.

Saenger, W. (1984). *Principles of Nucleic Acid Structure*. Springer-Verlag, Berlin.
Dated in many areas, but still an important reference book, providing a wealth of background information and detail, and historic perspective.

Sinden, R. R. (1994). *DNA Structure and Function*. Academic Press, San Diego.

Watson, J. D. (1968). *The Double Helix*. Weidenfeld and Nicolson, London.
A highly personalized and entertaining account of the discovery of DNA structure—essential reading for all budding students of DNA structure.

More detailed accounts of particular topics can be found in the following articles.

X-ray crystallography

Arnott, S. (1970). *Progress in Biophysics and Molecular Biology*, **21**, 265.
An account of the principles of nucleic acid fibre diffraction analysis.

Drenth, J. (1994). *Principles of Protein X-ray Crystallography*. Springer, New York.

McRee, D. E. (2000). *Practical Protein Crystallography*, 2nd edn, Academic Press, San Diego.

Rhodes, G. (1999). *Crystallography Made Crystal Clear*, 2nd edn, Academic Press, San Diego. Much relevant information can also be found on; http://www.usm.maine.edu/~rhodes/CMCC/index.html.

Rossmann, M. G. and Arnold, E. (ed.) (2001). *International Tables for X-ray Crystallography, Volume F, Crystallography of Biological Macromolecules*. Kluwer, Dordrecht.

NMR methods

Lane, A. N. (1994). *Methods in Enzymology*, **261**, 413.

Patel, D. J., Shapiro, L., and Hare, D. R. (1987). *Quarterly Reviews of Biophysics*, **20**, 1.

Roberts, G. C. K. (ed.) (1993). *NMR of Macromolecules: a Practical Approach*. Oxford University Press, Oxford.

Schmitz, U. and James, T. L. (1995). *Methods in Enzymology*, **261**, 1.

Wemmer, D. E. (1991). *Current Opinion in Structural Biology*, **1**, 452.

Wüthrich, K. (1986). *NMR of Proteins and Nucleic Acids*. John Wiley, New York.

Molecular modelling and simulation methods

Beveridge, D. L. and McConnell, K. J. (2000). *Current Opinion in Structural Biology*, **10**, 182–96.

Goodfellow, J. M. and Williams, M. A. (1992). *Current Opinion in Structural Biology*, **2**, 211.

McCammon, J. A. and Harvey, S. C. (1987). *Dynamics of Proteins and Nucleic Acids*. Cambridge University Press, Cambridge, UK.

Tsui, V. and Case, D. A. (2000). *Journal of the American Chemical Society*, **122**, 2489.

van Gunsteren, W. F. and Berendsen, H. J. C. (1990). *Angewante Chemie, International Edition*, **29**, 992.

Chemical and enzymatic probes of DNA structure

Fox, K. R. (ed.) (1997). *Drug-DNA Interaction Protocols*. Humana Press, New Jersey.

Tullius, T. D. (1989). In *Nucleic Acids and Molecular Biology* (Eckstein, F. and Lilley, D. M. J., eds), Volume 3, pp. 1–12. Springer-Verlag, Berlin.

Tullius, T. D. (1991). *Current Opinion in Structural Biology*, **1**, 428.

The nucleic acid database

Berman, H. M., Olson, W. K., Beveridge, D. L., Westbrook, J., Gelbin, A., Demeny, T. *et al.* (1992). *Biophysical Journal*, **63**, 751.

Berman, H. M., Gelbin, A., and Westbrook, J. (1996). *Progress in Biophysics and Molecular Biology*, **66**, 255.

Some useful internet sites

http://www.rcsb.org   The site for the macromolecular (protein and nucleic acid) structure database.

http://www.ndb.rutgers.edu   The site for the Nucleic Acid Database.

Other NDB sites are:

http://www.ndb.icr.ac.uk   The Institute of Cancer Research, UK,

http://ndb.sdsc.edu/NDB/   San Diego Supercomputer Centre, USA,

http://ndbserver.nibh.go.jp/NDB/   Structural Biology Centre, NIBH, Japan,

http://bmrd.wisc.edu   The BioMagResBank Repository of NMR Structures.

http://www.umass.edu/microbiol/rasmol, http://www.bernstein-plus-sons/software/rasmol Two sites for the freeware molecular display program RasMol, which can run on a wide variety of workstations, PCs and Apple computers.

http://cns.csb.yale.edu   The site for the CNS software for crystallographic and NMR structure refinements.

2

The building-blocks of DNA and RNA

2.1 Introduction

Chemical degradation studies, in the early years of this century, on material extracted from cell nuclei established that the high molecular-weight 'nucleic acid' was actually composed of individual acid units, termed nucleotides. Four distinct types were isolated—guanylic, adenylic, cytidylic, and thymidylic acids. These could be further cleaved to phosphate groups and four distinct nucleosides. The latter were subsequently identified as consisting of a deoxypentose sugar and one of four nitrogen-containing heterocyclic bases. Thus, each repeating unit in a nucleic acid polymer comprises these three units linked together— a phosphate group, a sugar, and one of the four bases.

The bases are planar aromatic heterocyclic molecules and are divided into two groups— the pyrimidine bases, thymine and cytosine and the purine bases, adenine and guanine. Their major tautomeric forms are shown in Fig. 2.1. Thymine is replaced by uracil in ribonucleic acids, which also have an extra hydroxyl group at the $2'$ position of their (ribose) sugar groups. The standard nomenclature for the atoms in nucleic acids, as approved by the International Union of Biochemistry, is shown in Figs 2.1. and 2.2. Accurate bond length and angle geometries for all bases, nucleosides and nucleotides have been well established by X-ray crystallographic analyses. The most recent surveys[1,2] have calculated mean values for these parameters (which define their equilibrium values) from the most reliable structures in the Cambridge and Nucleic Acid Databases. These have been incorporated in several implementations of the AMBER and CHARMM force-fields widely used in molecular mechanics and dynamics modelling, and in a number of computer packages for both crystallographic and NMR structural analyses.[3] Accurate crystallographic analyses, at very high resolution, can also directly yield quantitative information on the electron-density distribution in a molecule, and hence on individual partial atomic charges. These charges for nucleosides have hitherto been obtained by *ab initio* quantum mechanical calculations, but are now available experimentally for all four DNA nucleosides.[4]

Individual nucleoside units are joined together in a nucleic acid in a linear manner, through phosphate groups attached to the $3'$ and $5'$ positions of the sugars (Fig. 2.2). Hence, the full repeating unit in a nucleic acid is a $3',5'$-nucleo*tide*.

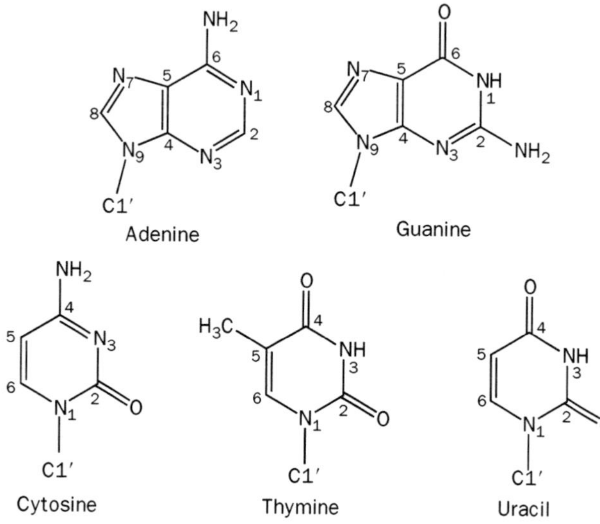

Fig. 2.1 The five bases of DNA and RNA.

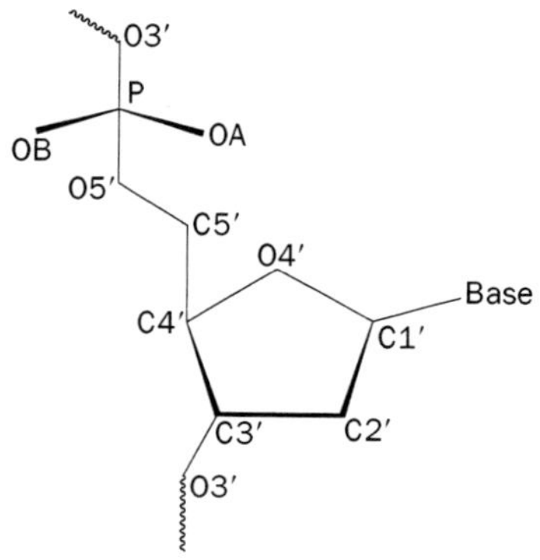

Fig. 2.2 The organization of repeating units in a polynucleotide chain.

Nucleic acid and oligonucleotide sequences use single-letter codes for the five unit nucleotides—A, T, G, C, and U. The two classes of bases can be abbreviated as Y (pyrimidine) and R (purine). Phosphate groups are usually designated as p. A single oligonucleotide chain is conventionally numbered from the 5′ end, for example, ApGpCpTpTpG has the 5′ terminal adenosine nucleoside, with a free hydroxyl at its 5′ position and thus the 3′ end guanosine has a free 3′ terminal hydroxyl group. The letter p, to denote intervening phosphate groups, is sometimes omitted when a sequence is written down. Chain direction is sometimes emphasized with 5′ and 3′ labels. Thus an antiparallel double-helical sequence can be written as:

5′CpGpCpGpApApTpTpCpGpCpG

3′GpCpGpCpTpTpApApGpCpGpC

or simply as

CGCGAATTCGCG

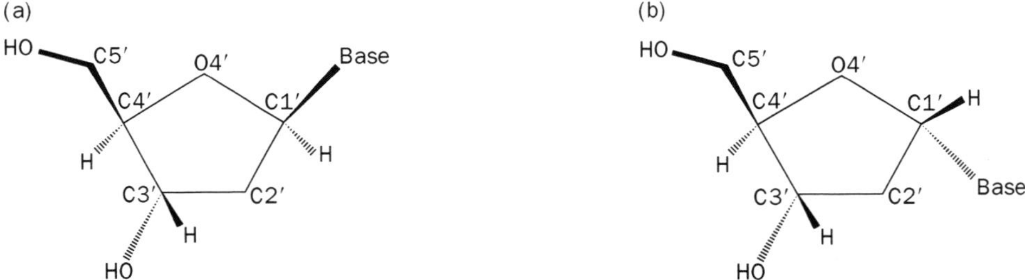

Fig. 2.3 (a) The stereochemistry of a natural β-nucleoside. Solid bonds are coming out of the plane of the page, towards the reader. Dashed bonds are going away from the reader. (b) The stereochemistry of an α-nucleoside.

Fig. 2.4 (a) A•T and (b) G•C base pairs, showing Watson–Crick hydrogen bonding.

Structural publications on DNA usually prefix a sequence with 'd', as in d(CGAT), to emphasize that the oligonucleotide is a deoxyribose one rather than being an oligoribonucleotide. The prefix 'r' to denote ribonucleosides, ribonucleotides, and their oligomers, is often used.

The bond between sugar and base is known as the glycosidic bond. Its stereochemistry is important. In natural nucleic acids the glycosidic bond is always β, that is the base is above the plane of the sugar when viewed onto the plane and therefore on the same face of the plane as the 5′ hydroxyl substituent (Fig. 2.3a). The absolute stereochemistry of other substituent groups on the deoxyribose sugar ring of DNA is defined such that when viewed end-on with the sugar ring oxygen atom O4′ at the rear (Fig. 2.4a), the hydroxyl group at the 3′ position is below the ring and the hydroxymethyl group at the 4′ position is above it. A unit nucleotide can have its phosphate group attached either at the 3′ or 5′ ends, and is thus termed either a 3′ or a 5′ nucleotide. It is chemically possible to construct α-nucleosides and from them α-oligonucleosides, which have their bases in the 'below' configuration relative to the sugar rings and their other substituents (Fig. 2.3b). These are much more resistant to nuclease attack than standard natural β-oligomers and have been used as antisense oligomers to mRNAs on account of their superior intracellular stability.

2.2 **Base pairing**

The realization that the planar bases can associate in particular ways by means of hydrogen bonding was a crucial step in the elucidation of the structure of DNA. The important early experimental data of Chargaff showed that the molar ratios of adenine : thymine and

Table 2.1 Hydrogen-bond distances (Å) in Watson–Crick base pairs.[6,7] Estimated standard deviations are in parentheses

U•A	N3–H $\cdots$ N1	2.835(8)
	O4 $\cdots$ H–N6	2.940(8)
C•G	O2 $\cdots$ H–N2	2.86(1)
	N3 $\cdots$ H–N1	2.95(1)
	N4–H $\cdots$ O6	2.91(1)

cytosine : guanine in DNA were both unity. This led to the proposal by Crick and Watson that in each of these pairs the purine and pyrimidine bases are held together by specific hydrogen bonds, to form planar base pairs. In native, double helical DNA the two bases in a base pair necessarily arise from two separate strands of DNA (with intermolecular hydrogen bonds) and so hold the DNA double helix together.[5]

The adenine : thymine (A•T) base pair has two hydrogen bonds compared to the three in a guanine : cytosine (G•C) one (Fig. 2.4). Fundamental to the Watson–Crick arrangement is that the sugar groups are both attached to the bases on the same side of the base pair. As will be seen in Chapter 3, this defines the mutual positions of the two sugar-phosphate strands in DNA itself. The two base pairs are required to be almost identical in dimensions by the Watson–Crick model. High resolution (0.8–0.9 Å) X-ray crystallographic analyses of the ribodinucleoside monophosphate duplexes r(GpC) and r(ApU) by A. Rich and colleagues in the early 1970s[6,7] has established accurate geometries for these A•T and G•C base pairs (Table 2.1). These structure determinations showed that there are only small differences in size between the two types of base pairings, as indicated by the distance between glycosidic carbon atoms in a base pair. The C1$'$ $\cdots$ C1$'$ distance in the G•C base pair structure is 10.67 Å, and 10.48 Å in the A•U-containing dinucleoside.

Theoretical studies on base pairing have enabled reliable estimates to be made of the extra stability conferred on a G•C base pair by the third hydrogen bond. Molecular dynamics simulations of both base pairs in an aqueous environment have given the free energies of A•T and G•C base pairs as −4.3 and −5.8 kcal/mol.[8] An alternative, reversed Watson–Crick A•T base pairs involving a 180° rotation of one base, had similar stability. Other A•T base pair arrangements have been predicted by theory[9] to be more stable than Watson–Crick pairing. All such alternatives would require significant changes in backbone conformation in order to be accommodated within a duplex. It is striking that these alternatives have not been observed in normal duplex DNA, suggesting that the requirements for optimal base stacking and minimal backbone distortions greatly favour the standard Watson–Crick arrangement.

2.3 Base and base pair flexibility

The individual bases in a nucleic acid are flat, but base pairs (and consecutive bases on an individual strand), can show considerable flexibility. This flexibility is to some extent dependent on the nature of the bases and base pairs themselves, but is more related to their base-stacking environments. Thus, descriptions of base morphology have become important in describing and understanding many sequence-dependent features and deformations of nucleic acids. The former features are often considered primarily at the dinucleoside local level, whereas longer range effects such as helix bending, can also be analysed at a more global level.

A number of rotational and translational parameters have been devised to describe these geometric relations between bases and base pairs, which were originally defined[10] in 1989 (the 'Cambridge Accord'). These definitions, together with the Cambridge Accord sign conventions, are given below. Unfortunately, there is now some confusion in the literature regarding these parameters, in part because two distinct types of approaches have been developed to calculate them[11]—the Cambridge Accord did not define a single unambiguous convention for their calculation. In one approach, the parameters are defined with respect to a global helical axis, which need not be linear. The other uses a set of local axes, one per dinucleotide step. Also, a variety of definitions of local and global axes have been used. The overall effect for most undistorted structures is fortunately that only a minority of parameters appear to have very different values depending on the method of calculation, using a number of the widely available programs (see below). Before detailing these, we introduce the parameters themselves. The major and minor grooves are the indentations in nucleic acid double helices formed as a consequence of the asymmetry of base pairs. The C1′-N9 (purine) and C1′-N1 (pyrimidine) base-sugar bonds are by convention on the minor-groove side, so that the C6/N7 (purine) and C4 (pyrimidine) base atoms and their substituents are on the major-groove side. The grooves and their characteristics are described further in Chapter 3.

(i) For individual base pairs

(a) **Propeller twist** (ω) between bases is the dihedral angle between normals to the bases, when viewed along the long axis of the base pair (Fig. 2.5). The angle has a negative sign under normal circumstances, with a clockwise rotation of the nearer base when viewed down the long axis. The long axis for a purine–pyrimidine base pair is defined as the vector between the C8 atom of the purine and the C6 of a pyrimidine in a Watson–Crick base pair. Analogous definitions can be applied to other non-standard base pairings in a duplex including purine– purine and pyrimidine–pyrimidine ones.

(b) **Buckle** (κ) is the dihedral angle between bases, along their short axis, after propeller twist has been set to 0° (Fig. 2.6). The sign of buckle is defined as positive if the

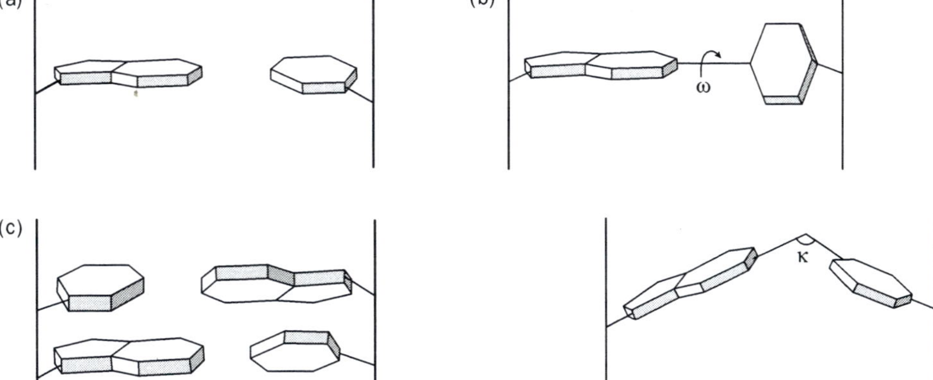

Fig. 2.5 Schematic views of a base pair with (a) zero and (b) high propeller twist. Panel (c) shows the effects of propeller twist on two successive base pairs.

Fig. 2.6 A schematic view of base pair buckle.

distortion is convex in the direction $5' \rightarrow 3'$ of strand 1. The change in buckle for succeeding steps, termed **cup**, has been found to be a useful measure of changes along a sequence. Cup is defined as the difference between the buckle at a given step, and that of the preceding one.

(c) **Inclination** (η) is the angle between the long axis of a base pair and a plane perpendicular to the helix axis. This angle is defined as positive for right-handed rotation about a vector from the helix axis towards the major groove.

(d) X **and** Y **displacements** define translations of a base pair within its mean plane in terms of the distance of the midpoint of the base pair long axis from the helix axis. X displacement is towards the major groove direction, when it has a positive value. Y displacement is orthogonal to this, and is positive if towards the first nucleic acid strand of the duplex.

(ii) For base pair steps

(e) **Helical twist** (Ω) is the angle between successive base pairs, measured as the change in orientation of the $C1'-C1'$ vectors on going from one base pair to the next, projected down the helix axis (Fig. 2.7). For an exactly repetitious double helix, helical twist is $360°/n$, where n is the unit repeat defined above.

(f) **Roll** (ρ) is the dihedral angle for rotation of one base pair with respect to its neighbour, about the long axis of the base pair. A positive roll angle opens up a base pair step towards the minor groove (Fig. 2.8). **Tilt** (τ) is the corresponding dihedral angle along the short (i.e. x-axis) of the base pair.

(g) **Slide** is the relative displacement of one base pair compared to another, in the direction of nucleic acid strand one (i.e. the Y displacement), measured between the midpoints of each C6–C8 base pair long axis.

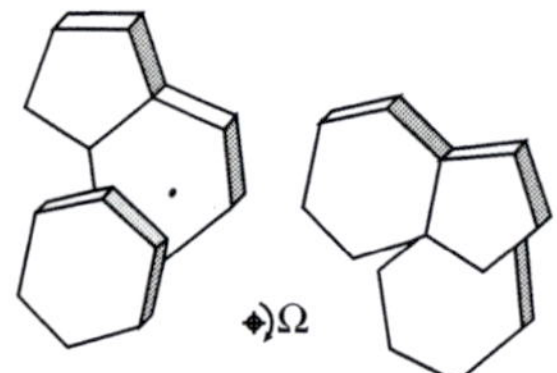

Fig. 2.7 View down two successive base pairs, showing the helical twist angle between them.

A standard coordinate reference frame for the calculation of these parameters has been proposed,[12] and has been endorsed by the successor to the Cambridge Accord, the 1999 Tsukuba Accord. Details of this are also available in *Journal of Molecular Biology* (2001), **313**, 229. The reference frame proposed is unambiguous and has the advantage of being able to produce values for the majority of local base pair and base step parameters

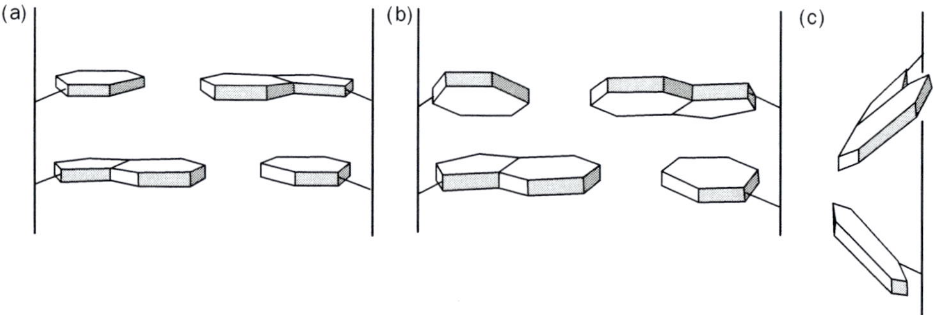

Fig. 2.8 Views of two successive base pairs, (a) with 0° roll angle between them, and (b) with a positive roll angle. (c) Shows a view of positive roll along the long axis of the base pairs.

which are almost independent of the algorithm used. A notable exception is rise, which is especially sensitive to the definition of origin and to small changes in buckle and roll. The right-handed reference frame used is shown in Fig. 2.9. It has the x-axis directed towards the major groove along the pseudo two-fold axis of an idealized Watson–Crick base pair (shown as •). The y-axis is along the long axis of the base pair, parallel to the $C1' \cdots C1'$ vector. The position of the origin is clearly dependent on the geometry of the bases and the base pair. These have been taken from the published compilations.[1,2]

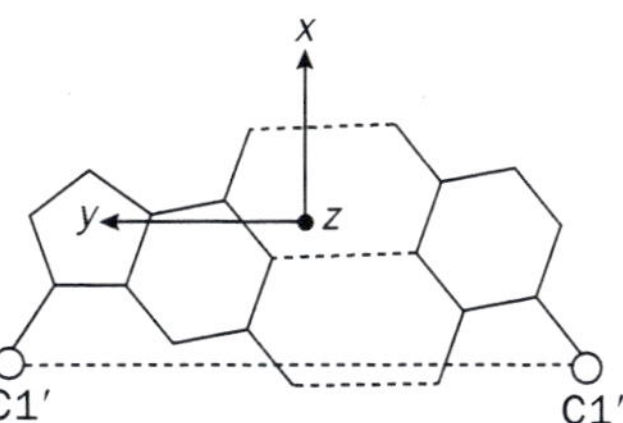

Fig. 2.9 Reference frame for idealized helical DNA, showing the axis origin at (•), along the pseudo two-fold axis between two successive base pairs.

The most widely used programs for base and base pair morphology calculations, all of which use the above parameter definitions, are

1. CURVES,[13,14] which calculates an optimized global helix axis which can be (and invariably is), curved. This is especially useful for irregular structures. The reference frame is defined at the base rather than the base pair level. Local parameters can also be obtained with this program, and irregular non-standard base pairing arrangements can be accommodated.
2. FREEHELIX, the successor to the original helical parameter program NEWHELIX, calculates an optimum linear helical axis, so that the derived parameters are basically global ones.[15] Local bending can also be calculated.[16]
3. RNA[17] uses the reference frame subsequently adopted by the Tsukuba Accord, and calculates local helical parameters. It can also be used for non-standard bases.
4. CEHS[18] also focuses on local parameters, and like FREEHELIX, uses the $C6 \cdots C8$ base pair vector as the y-axis for the reference frame. Roll and tilt are commuted into a single variable RollTilt (Γ).

Readers of crystallographic or NMR structure analysis literature should be aware of whether local or global axis definitions have been used in parameter calculation. Ambiguities can be clarified by use of the NDB, which can produce tables of helical parameters calculated with all of these methods.

2.4 Sugar puckers

The five-membered deoxyribose sugar ring in DNA is inherently non-planar. This non-planarity is termed puckering. The precise conformation of a deoxyribose ring can be completely specified by the five endocyclic torsion angles within it (Fig. 2.10). The ring puckering arises from the effect of non-bonded interactions between substituents at the four ring carbon atoms— the energetically most stable conformation for the ring has all substituents as far apart as possible. Thus,

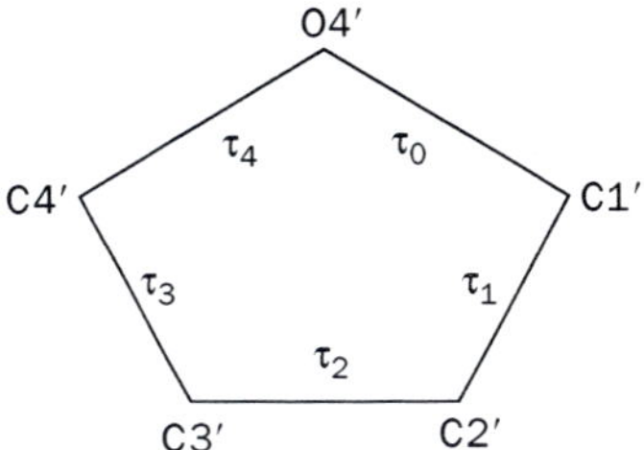

Fig. 2.10 The five internal torsion angles in a ribose ring.

different substituent atoms would be expected to produce differing types of puckering. The puckering can be described by either

(1) a simple qualitative description of the conformation in terms of atoms deviating from ring coplanarity; or
(2) precise descriptions in terms of the ring internal torsion angles.

In principle, there is a continuum of interconvertible puckers, separated by energy barriers. These various puckers are produced by systematic changes in the ring torsion angles. The puckers can be succinctly defined by the parameters P and τ_m.[19] The value of P, the phase angle of pseudorotation, indicates the type of pucker since P is defined in terms of the five torsion angles τ_0–τ_4:

$$\tan P = \frac{(\tau_4 + \tau_1) - (\tau_3 + \tau_0)}{2 * \tau_2 * (\sin 36° + \sin 72°)}$$

and the maximum degree of pucker, τ_m, by

$$\tau_m = \frac{\tau_2}{\cos P}$$

The pseudorotation phase angle can take any value between $0°$ and $360°$. If τ_2 has a negative value, then $180°$ is added to the value of P. The pseudorotation phase angle is commonly represented by the pseudorotation wheel, which indicates the continuum of ring puckers (Fig. 2.11). Values of τ_m indicate the degree of puckering of the ring; typical experimental values from crystallographic studies on mononucleosides are in the range 25–45°. The five internal torsion angles are not independent of each other, and so to a good approximation any one angle τ_j can be represented in terms of just two variables:

$$\tau_j = \tau_m \cos [P + 0.8\pi(j - 2)]$$

A large number of distinct deoxyribose ring pucker geometries have been observed experimentally, by X-ray crystallography and NMR techniques. When one ring atom is out of the plane of the other four, the pucker type is an envelope one. More commonly, two atoms deviate from the plane of the other three, with these two either side of the plane. It is usual for one of the two atoms to have a larger deviation from the plane than the other,

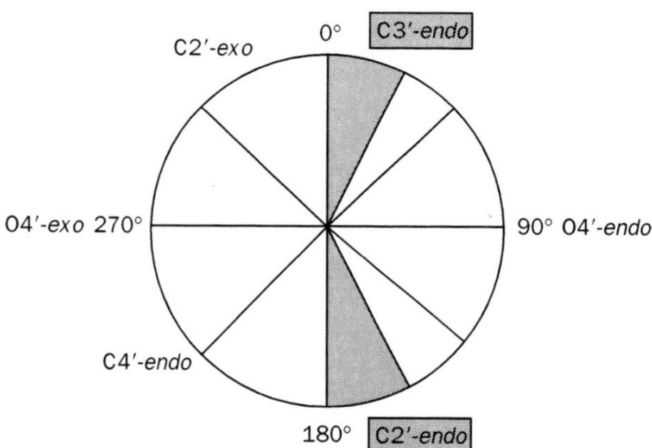

Fig. 2.11 The pseudorotation wheel for a deoxyribose sugar. The shaded areas indicate the preferred ranges of the pseudorotation angle for the two principal sugar conformations.

resulting in a twist conformation. The direction of atomic displacement from the plane is important. If the major deviation is on the same side as the base and C4′–C5′ bond, then the atom involved is termed *endo*. If it is on the opposite side, it is called *exo*. The most commonly observed puckers in crystal structures of isolated nucleosides and nucleotides are either close to C2′-*endo* or C3′-*endo* types (Figs 2.12a,b). In practice, these pure envelope forms are rarely observed, largely because of the differing substituents on the ring. Consequently, the puckers are then best described in terms of twist conformations. When the major out-of-plane deviation is on the *endo* side, there is a minor deviation on the opposite, *exo* side. The convention used for describing a twist deoxyribose conformation is that the major out-of-plane deviation is followed by the minor one, for example C2′-*endo*, C3′-*exo*. The C2′-*endo* family of puckers have *P* values in the range 140–185°; in view of their position on the pseudorotation wheel, they are sometimes termed S (south) conformations. The C3′-*endo* domain has *P* values in the range −10 to +40°, and its conformation is termed N (north).

The pseudorotation wheel implies that deoxyribose puckers are free to interconvert. In practice, there are energy barriers between major forms. The exact size of these barriers has been the subject of considerable study.[20,21] The consensus is that the barrier height is dependent on the route around the pseudorotation wheel. For interconversion of C2′-*endo* to C3′-*endo* the preferred pathway is via the O4′-*endo* state, with a barrier of 2–5 kcal/mol found from an analysis of a large body of experimental data,[20] and a somewhat smaller (potential energy) value of 1.5 kcal/mol from a molecular dynamics study.[22] The former value, being an experimental one, represents the total free energy for interconversion.

Relative populations of puckers can be monitored directly by NMR measurements of the ratio of coupling constants between H1′–H2′ and H3′–H4′ protons. These show that in contrast to the 'frozen-out' puckers found in the solid state structures of nucleosides

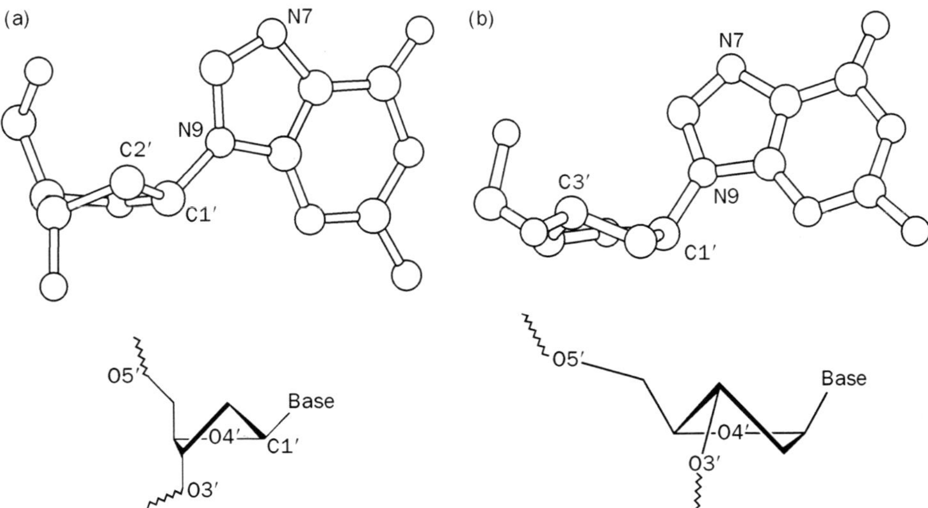

Fig. 2.12 (a) C2′-*endo* sugar puckering for a guanosine nucleoside, viewed along the plane of the sugar ring. (b) C3′-*endo* sugar puckering for a guanosine nucleoside, with the sugar viewed in the same direction as in panel (a).

and nucleotides, there is rapid interconversion in solution. Nonetheless, the relative populations of the major puckers are dependent on the type of base attached. Purines show a preference for the C2′-*endo* pucker conformational type whereas pyrimidines favour C3′-*endo*. Deoxyribose nucleosides are primarily (>60 per cent) in the C2′-*endo* form and ribonucleosides favour C3′-*endo*. The latter are significantly more restricted in their mobility; this has importance for the structures of oligoribonucleotides (see Chapter 6). These differences in puckering equilibria and hence in their relative populations in solution and in molecular dynamics simulations, are reflected in the patterns of puckers found in surveys of crystal structures.[23] Again, this is a demonstration of the complementarity of information provided by the different structural techniques. Sugar pucker preferences have their origin in the non-bonded interactions between substituents on the sugar ring, and to some extent on their electronic characteristics. For example, the C3′-*endo* pucker (Fig. 2.7b) would have hydroxyl substituents at the 2′ and 3′ positions further apart than with C2′-*endo* pucker; hence the preference of the former by ribonucleosides.

Correlations have been found from numerous crystallographic and NMR studies, between sugar pucker and several backbone conformational variables, both in isolated nucleosides/nucleotides and in oligonucleotide structures. These are discussed later in this chapter. Changes in sugar pucker are important in oligo- and polynucleotides because they can alter the orientation of C1′, C3′, and C4′ substituents, resulting in major changes in backbone conformation and overall structure, as indeed is found (Chapter 3). Sugar pucker is thus an important determinant of oligo- and polynucleotide conformation.

2.5 Conformations about the glycosidic bond

The glycosidic bond links a deoxyribose sugar and a base, being the C1′–N9 bond for purines and the C1′–N1 bond for pyrimidines. The torsion angle χ around this single bond can in principle adopt a wide range of values, although as will be seen, structural constraints result in marked preferences being observed. Glycosidic torsion angles are defined in terms of the four atoms:

O4′–C1′–N9–C4 for purines

O4′–C1′–N1–C2 for pyrimidines

Theory has predicted two principal low-energy domains for the glycosidic angle, in accord with experimental findings for a large number of nucleosides and nucleotides. The *anti* conformation has the N1, C2 face of purines and the C2, N3 face of pyrimidines directed away from the sugar ring (Fig. 2.13a) so that the hydrogen atoms attached to C8 of purines and C6 of pyrimidines, are lying over the sugar ring. Thus, the Watson–Crick hydrogen-bonding groups of the bases are directed away from the sugar ring. These orientations are reversed for the *syn* conformation, with these hydrogen-bonding groups now oriented towards the sugar and especially its O5′ atom (Fig. 2.13b). A number of crystal structures of *syn* purine nucleosides have observed hydrogen bonding between the O5′-atom and the N3 base atom, which would stabilize this conformation. Otherwise, for purines, the *syn* conformation is slightly less preferred than the *anti*, on the basis of fewer non-bonded steric clashes in the latter case. The principal exceptions to this rule are guanosine-containing nucleotides, which have a small preference for the *syn* form because of favourable electrostatic interactions between the exocyclic N2 amino group of guanine and the 5′-phosphate atom. For pyrimidine nucleotides, the *anti* conformation is

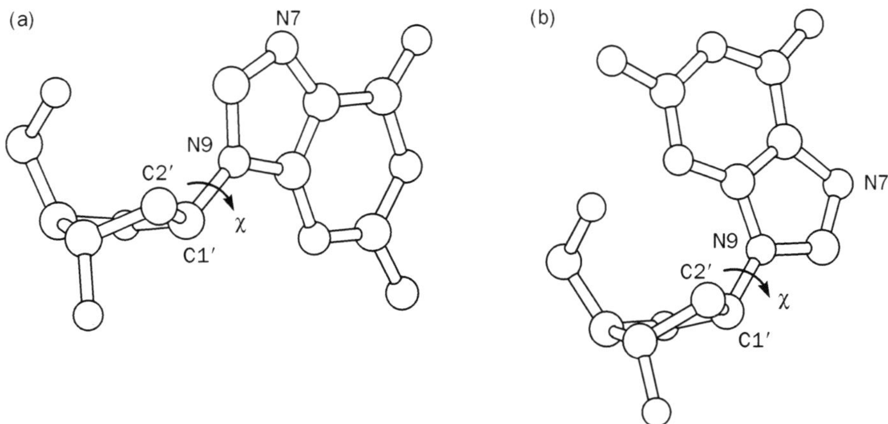

Fig. 2.13 (a) A guanosine nucleoside with the glycosidic angle χ set in an *anti* conformation. (b) Guanosine now in the *syn* conformation.

Table 2.2 Energy differences (kcal/mol) between glycosidic angle conformers for B-form DNA nucleotides, calculated with the AMBER program

Nucleotides	Energy difference
A(*anti*) > A(*syn*)	0.3
T(*anti*) > T(*syn*)	1.7
G(*syn*) > G(*anti*)	3.3
C(*anti*) > C(*syn*)	1.8

preferred over the *syn*, because of unfavourable contacts between the O2 oxygen atom of the base and the 5′-phosphate group. The results of molecular-mechanics energy minimizations on all four DNA nucleotides in both *syn* and *anti* forms (using the AMBER all-atom force field) are fully in accord with these observations (Table 2.2).

The sterically preferred ranges for the two domains of glycosidic angles are:

Anti: $-120 > \chi > 180°$

Syn: $0 < \chi < 90°$

Values of χ in the region of about $-90°$ are often described as 'high *anti*'. There are pronounced correlations between sugar pucker and glycosidic angle, which reflect the changes in non-bonded clashes produced by C2′-*endo* versus C3′-*endo* puckers. Thus, *syn* glycosidic angles are not found with C3′-*endo* puckers due to steric clashes between the base and the H3′ atom, which points towards the base in this pucker mode.

2.6 The backbone torsion angles and correlated flexibility

The phosphodiester backbone of an oligonucleotide has six variable torsion angles (Fig. 2.14), designated $\alpha, \ldots, \zeta$, in addition to the five internal sugar torsions $\tau_0, \ldots, \tau_4$

and the glycosidic angle χ. As will be seen, a number of these have highly correlated values (and therefore correlated motions in a solution environment). Steric considerations alone dictate that the backbone angles are restricted to discrete ranges[24,25] (Fig. 2.15), and are accordingly not free to adopt any value between 0 and 360°. Figure 2.15 uses a conformational wheel to show these preferred values, which are directly readable from their positions around the wheel. The fact that angles α, β, γ, and ζ each have three allowed ranges, together with the broad range for angle ε that includes two staggered regions, leads to a large number of possible low-energy conformations for the unit nucleotide, especially

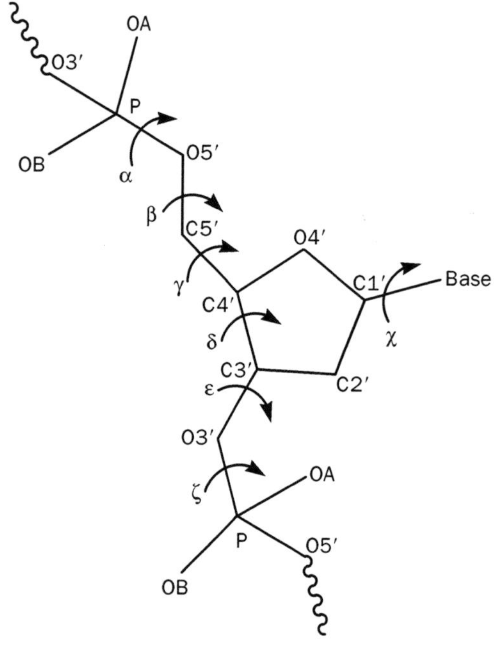

Fig. 2.14 The backbone torsion angles in a unit nucleotide. Each rotatable bond is indicated by a curved arrow.

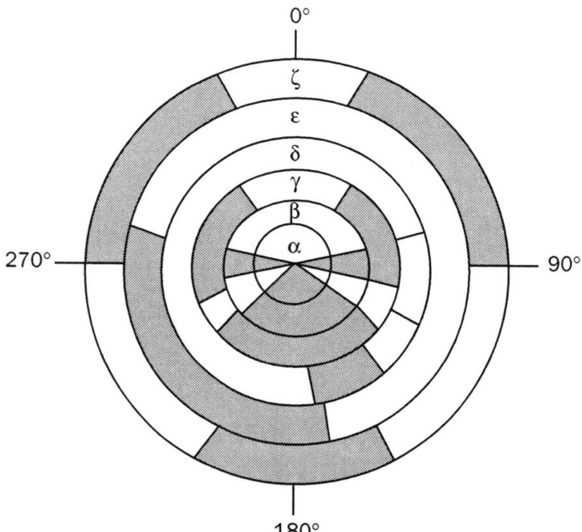

Fig. 2.15 Conformational wheel showing the allowed ranges of backbone torsion angles (shaded) in nucleosides, nucleotides, deoxyoligonucleotides and deoxypolynucleotides.

when glycosidic angle and sugar pucker flexibility is taken into account. In reality, only a small number of DNA oligonucleotide and polynucleotide structural classes have actually been observed out of this large range of possibilities; this is doubtless in large part due to the restraints imposed by Watson–Crick base pairing on the backbone conformations when two DNA strands are intertwined together. In contrast, crystallographic and NMR studies on a large number of standard and modified mononucleosides and nucleotides have shown their considerably greater conformational diversity, in accord with the possibilities indicated in Fig. 2.15. For these, backbone conformations in the solid state and in solution are not always in agreement; the requirements for efficient packing in the crystal can often overcome the modest energy barriers between different values for a particular torsion angle. Many large RNA molecules are characterized by a wide range of base . . . base interactions, which can therefore adopt a variety of backbone conformations.

A common convention for describing these backbone angles is to term values of ~ 60° as *gauche*$^+$(g$^+$), $-60°$ as *gauche*$^-$ (g$^-$) and ~180° as *trans* (t). Thus, for example, angles α (about the P–O5′ bond) and γ (the exocyclic angle about the C4′–C5′ bond), can be in the g$^+$, g$^-$ or t conformations. The two torsion angles around the phosphate group itself, α and ζ, have been found to show a high degree of flexibility in various dinucleoside crystal structures, with the tg$^-$, g$^-$g$^-$ and g$^+$g$^+$ conformations all having been observed.[26] As will be described in Chapter 3, only the g$^-$g$^-$ conformation can place successive nucleotide units in arrangements that have their bases in potential hydrogen-bonding positions with respect to a second nucleotide strand. Thus, this is the phosphate conformation for DNA and RNA double helices. The torsion angle β, about the O5′–C5′ bond, is almost always found to be *trans*. All three possibilities for the γ angle have been observed in nucleoside crystal structures, although the g$^+$ conformation predominates in right-handed oligo- and polynucleotide double helices. The *trans* conformation for γ places the 5′-phosphate group in quite a distinct position with respect to the deoxyribose ring. The torsion angle δ around the C4′–C3′ bond adopts values that relate to the pucker of the sugar ring, since the internal ring torsion angle τ_3, (also around this bond), has a value of about 35° for C2′-*endo* and about 40° for C3′-*endo* puckers. δ is about 75° for C3′-*endo* and about 150° for C2′-*endo* puckers.

An alternative nomenclature for torsion angle ranges used by some workers in the nucleic acid field is that common in organic chemistry (the Klyne-Prelog system). In this, the *syn* (s) designation is given to angles clustered around 0°, and *anti* (a) for those around 180°. Intermediate angles are defined as $\pm$*synclinal* ($\pm sc$) for around $\pm60°$, and $\pm$*anticlinal* ($\pm ac$) for around $\pm120°$.

There are a number of well-established correlations involving pairs of these backbone torsion angles, as well as sugar pucker and glycosidic angle. These have been observed in mononucleosides and nucleotides (which are inherently more flexible in solution as well as being more subject to packing forces in the crystal), and more recently, in oligonucleotides.[27,28] The existence of such correlations is important: it means that the atomic motions in oligo- and polynucleotides follow concerted patterns of inter-dependence. In general, these correlations are due to the diminution of non-bonded contacts that occur with particular conformations. Some of the more important correlations that have been observed in mononucleosides and nucleotides are:

1. Between sugar pucker and glycosidic angle χ, especially for pyrimidine nucleosides. C3′-*endo* pucker is usually associated with median-value *anti* glycosidic angles, whereas

C2'-*endo* puckers are commonly found with high *anti* χ angles. *Syn* glycosidic angle conformations show a marked preference for C2'-*endo* sugar puckers.

2. The C4'–C5' torsion angle γ is correlated with the glycosidic angle and to some extent with sugar pucker and backbone angle α.[27] *Anti* glycosidic angles tend to correlate with g$^+$ conformations for γ.

References

1 Clowney, L., Jain, S. C., Srinivasan, A. R., Westbrook, J., Olson, W. K., and Berman, H. M. (1996). *Journal of the American Chemical Society*, **118**, 509.

2 Gelbin, A., Schneider, B., Clowney, L., Hsieh, S.-H., Olson, W. K., and Berman, H. M. (1996). *Journal of the American Chemical Society*, **118**, 519.

3 Parkinson, G., Vojtechovsky, J., Clowney, L., Brünger, A. T., and Berman, H. M. (1996). *Acta Crystallographica*, **D52**, 57.

4 Pearlman, D. A. and Kim, S.-H. (1990). *Journal of Molecular Biology*, **211**, 171.

5 Watson, J. D. and Crick, F. H. C. (1953). *Nature*, **171**, 737.

6 Seeman, N. C., Rosenberg, J. M., Suddath, F. L., Kim, J. P., and Rich, A. (1976). *Journal of Molecular Biology*, **104**, 109–144.

7 Rosenberg, J. M., Seeman, N. C., Day, R. O., and Rich, A. (1976). *Journal of Molecular Biology*, **104**, 145.

8 Stofer, E., Chipot, C., and Lavery, R. (1999). *Journal of the American Chemical Society*, **121**, 9503.

9 Gould, I. R. and Kollman, P. A. (1994). *Journal of the American Chemical Society*, **116**, 2493.

10 (1989). *EMBO Journal*, **8**, 1.

11 Lu, X.-J., Babcock, M. S., and Olson, W. K. (1999). *Journal of Biomolecular Structure and Dynamics*, **16**, 833.

12 Lu, X.-J. and Olson, W. K. (1999). *Journal of Molecular Biology*, **285**, 1563.

13 Lavery, R. and Sklenar, H. (1988). *Journal of Biomolecular Structure and Dynamics*, **6**, 63.

14 Lavery, R. and Sklenar, H. (1989). *Journal of Biomolecular Structure and Dynamics*, **6**, 655.

15 Dickerson, R. E. (1998). *Nucleic Acids Research*, **26**, 1906.

16 Goodsell, D. S. and Dickerson, R. E. (1994). *Nucleic Acids Research*, **22**, 5497.

17 Babcock, M. S, Pednault, E. P. D., and Olson, W. K. (1994). *Journal of Molecular Biology*, **237**, 125.

18 Lu, X.-J., El Hassan, M. A., and Hunter, C. A. (1997). *Journal of Molecular Biology*, **273**, 668.

19 Altona, C. and Sundaralingam, M. (1972). *Journal of the American Chemical Society*, **94**, 8205–8212.

20 Olson, W. K. and Sussman, J. L. (1982) *Journal of the American Chemical Society*, **104**, 270.

21 Olson, W. K. (1982) *Journal of the American Chemical Society*, **104**, 278

22 Harvey, S. C. and Prabhakaran, M. (1986). *Journal of the American Chemical Society*, **108**, 6128.

23 Murray-Rust, P. and Motherwell, S. (1978). *Acta Crystallographica*, **B34**, 2534.

24 Sundaralingam, M. (1969). *Biopolymers*, **7**, 821.

25 Olson, W.K. (1982) In *Topics in Nucleic Acid Structure, Part 2* (ed. Neidle, S.), pp. 1–79. Macmillan Press, London.

26 Kim, S.-H., Berman, H. M., Seeman, N. C., and Newton, M. D. (1973). *Acta Crystallographica*, **B29**, 703.

27 Schneider, B., Neidle, S., and Berman, H. M. (1997). *Biopolymers*, **42**, 113.

28 Packer, M. J. and Hunter, C. A. (1998). *Journal of Molecular Biology*, **280**, 407.

Further reading

Jeffrey, G. A. and Saenger, W. (1991). *Hydrogen Bonding in Biological Structures*. Chapters 15, 16. Springer-Verlag, Berlin.

Voet, D. and Rich, A. (1970) *Progress in Nucleic Acid Research and Molecular Biology*, **10**, 183.

3

DNA structure as observed in fibres and crystals

3.1 Structural fundamentals

3.1.1 Helical parameters

The fibre diffraction method for determining the structures of polymeric DNA (and RNA) molecules has been outlined in Chapter 1. Examination of X-ray diffraction photographs of oriented fibres obtained from them enables the basic parameters defining the dimensions of the repeating helix, to be calculated. Measurement of the spacing between the layer lines directly gives the reciprocal of the helical pitch P, defined as the distance, parallel to the helix axis, between successive nucleotide units per complete turn of the helix (Fig. 3.1). The presence of a meridional reflection for a fibre that is oriented perpendicular to the X-ray direction, indicates the presence of a structural periodicity in that direction. Analysis of the meridional reflection gives the regular repeat distance d (the helical rise if the nucleotide units are perpendicular to the helix axis). The unit repeat n, the number of nucleotide units in a single full turn of helix, that is per pitch P, is thus P/d. Helical rise is more generally the distance between successive nucleotide units, projected to be parallel to the helix axis if the units are not strictly perpendicular to this axis. The diameter of the helix can be derived from the dimensions of the closely packed unit cell projection down the fibre axis.

3.1.2 Base pair morphological features

Base pairs are not necessarily planar, and indeed are rarely so. This is in part because the geometric requirements of hydrogen bonding are not especially stringent, and in particular the angle subtended at the hydrogen atom (donor $\cdots$ H–acceptor) can deviate by up to about 35° from linearity without appreciable loss of hydrogen-bond energy. Other factors such as the need to avoid steric clash in some base pair $\cdots$ base pair non-bonded interactions within a helix, can result in distortions of base pairs from planarity, since these do not distort hydrogen bonds beyond these energetically favourable limits. These deviations from planarity can take place in a number of ways that can be categorized in two groups of local base pair morphological features,[1] as described in Chapter 2. The first category involves individual base pairs. It specifies both movements of one base

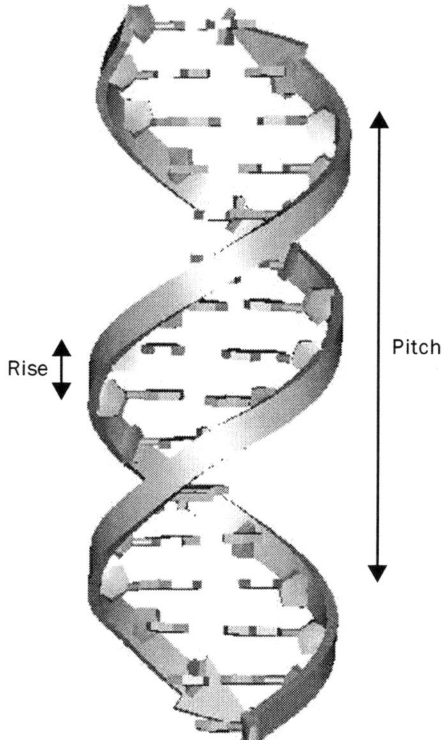

Fig. 3.1 Schematic DNA double helix, with backbones shown in ribbon representation.

relative to another within a base pair (e.g. propeller twist and buckle), and movement of the base pair relative to global axes in the helix. The second category is concerned with the relative movements of base pairs in successive base pairs (base pair steps), for example, helical twist and roll. It defines the relative relationships of successive base pairs in a structure. Inevitably the second set of features incorporate the first, so correlations between them can be expected.

3.2 Polynucleotide structures from fibre diffraction studies

3.2.1 Classic DNA structures

Information on the dimensions of polynucleotide double helices was initially derived from the diffraction patterns of DNA fibres at high (92 per cent) relative humidity, which showed a readily interpretable and characteristic Maltese cross pattern of strong diffraction intensities. Examination of this pattern, which arises from the B-DNA double helix, led directly to the original Watson–Crick model in 1953.[2] Subjecting DNA fibres to other conditions produce rather different diffraction patterns. In particular, lower (65–75 per cent) relative humidity conditions result in the A-DNA pattern, which typically has many more diffraction maxima, indicating greater crystallinity and hence order in the fibres. A number of other forms have subsequently been found (see Section 3.2.2), most of which are sub-classes of the A and B double helices, the two most important ones for random sequence DNA. Their exactly repetitious structures are often referred to as 'canonical DNA' ones. Double-helical DNA is thus highly polymorphic, the differing forms corresponding to distinct yet inter-convertible molecular structures. Some DNA structural types have only been found with defined-sequence polynucleotides rather than with random-sequence DNA (the standard A and B forms can occur with most sequences). Under appropriate experimental conditions, the diffraction pattern of such a polynucleotide is best fitted to a repeating unit, that rather than being a simple mono-nucleotide, can be a dinucleotide or even an oligonucleotide. Some of these repetitious fibres produce semi-crystalline diffraction patterns that give intensity data to significantly higher resolution than classic calf thymus DNA fibres. In these instances, the relatively high number of observed diffraction intensities, can result in structures that are as reliable as those of standard A and B polynucleotide helices.

B-DNA, the classic structure first described by Watson and Crick, has subsequently been refined[3] using the linked-atom least-squares procedure developed by Arnott and his group. The helical repeat is a single nucleotide unit, with necessarily all of the nucleotides in the structure having the conformation of this unit. Helical parameters for B- and other DNA polynucleotides are given in Table 3.1. The backbone conformation (Table 3.2) has high *anti* glycosidic angles and C2'-*endo* sugar puckers (Fig. 3.2).

Table 3.1 Selected helical parameters for various polymorphs of DNA polynucleotides, derived from fibre diffraction studies. Values are mostly from ref. (3)

	Unit repeat	Rise (Å)	Helical twist (°)	Base pair propeller twist (°)	Base step roll (°)	Base pair inclination (°)
A	11	2.54	32.7	−10.5	0.0	22.6
B	10	3.38	36.0	−15.1	0.0	2.8
C	28/3	3.31	38.6	−1.8	0.0	−8.2
D	8	3.01	45.0	−21.0	0.0	−13.0
Z (C)	6	7.25	−49.3	8.3	5.6	0.1
Z (G)	6	7.25	−10.3	8.3	−5.6	0.1

Table 3.2 Backbone conformational angles in (°) for various polymorphs of DNA polynucleotides, taken from ref. (3)

	α	β	γ	δ	ϵ	ζ	χ
A	−52	175	42	79	−148	−75	−157
B	−30	136	31	143	−141	−161	−98
C	−37	−160	37	157	161	−106	−97
D	−59	156	64	145	−163	−131	−102
Z (C)	−140	−137	51	138	−97	82	−154
Z (G)	52	179	−174	95	−104	−65	59

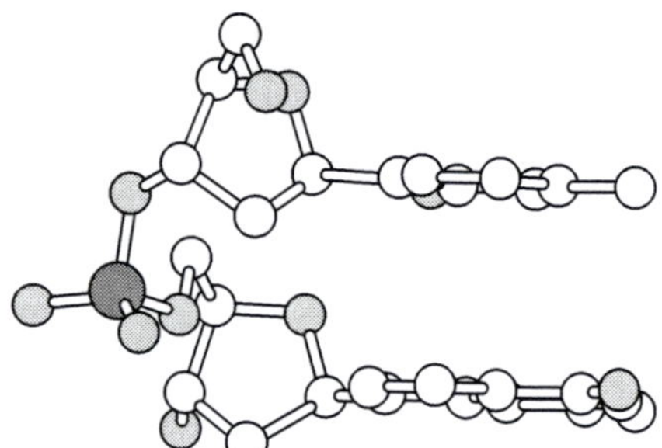

Fig. 3.2 Conformation of two successive nucleotides in one strand of canonical B-DNA, from fibre diffraction analysis.

(a)

(c)

(b)

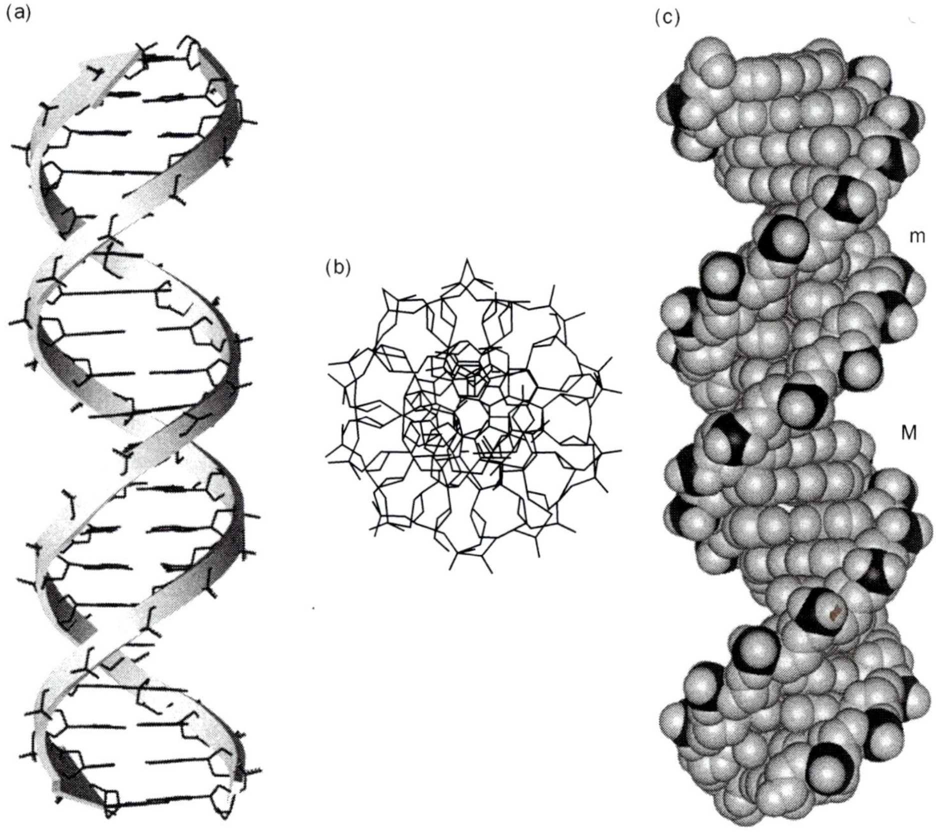

Fig. 3.3 (a) The structure of canonical B-DNA, from fibre diffraction analysis; (b) view down the helix axis; (c) space-filling representation of B-DNA, showing major and minor grooves. Phosphorus atoms are coloured black.

The right-handed double helix has 10 base pairs per complete turn, with the two polynucleotide chains wound anti-parallel to each other (Fig. 3.3a) and linked by Watson–Crick A•T and G•C base pairs. The paired bases are almost exactly perpendicular to the helix axis, and they are stacked over the axis itself (Fig. 3.3b). Consequently the base pair separation is the same as the helical rise—3.4 Å. An important consequence of the Watson–Crick base pairing arrangement is that the two deoxyribose sugars linked to an individual base pair are on the same side of it. So, when successive base pairs are stacked on each other in the helix, the gap between these sugars forms continuous indentations in the surface that wind along, parallel to the sugar–phosphodiester chains. These indentations are termed **grooves**. Groove width can be defined as the perpendicular distance between phosphate groups on opposite strands, minus the van der Waals diameter of a phosphate group (5.8 Å), although it is equally valid in some circumstances to use pairs of other atoms, such as C1′ or O4′. Groove depths are normally defined in terms of the differences in cylindrical polar radii between phosphorus and N2 guanine or N6 adenine atoms, for minor and major grooves respectively.

The asymmetry in the base pairs results in two parallel types of groove, whose dimensions (especially their depths) are related to the distances of base pairs from the axis of the helix and their orientation with respect to the axis. The B-DNA wide major groove (Fig. 3.3c and Table 3.3) is almost identical in depth to the much narrower minor groove, which has the hydrophobic hydrogen atoms of the sugar groups forming its walls. In general, the major groove is richer in base substituents—O6, N6 of purines and N4, O4 of pyrimidines compared to the minor one. This, together with the steric differences between the two, has important consequences for interaction with other molecules (see Chapters 4–7). Base pair and base step morphologies for several polynucleotide helices are given in Table 3.1.

The A-DNA duplex also has a single nucleotide unit as the helical repeat (Fig. 3.4). This has C3′-*endo* sugar puckers, which brings consecutive phosphate groups on the nucleotide chains closer together—5.9 Å compared to the 7.0 Å in B-form DNA, and alters the glycosidic angle from high *anti* to *anti*. As a consequence, the base pairs are

Table 3.3 Groove dimensions (Å) in polynucleotide DNA double helices, from fibre diffraction analyses (taken from ref. (3) for the A–D helices and from the single-crystal data (refs 72, 75) for the Z helix)

Polymorph	Major groove		Minor groove	
	Width	Depth	Width	Depth
A	2.2	13.0	11.1	2.6
B	11.6	8.5	6.0	8.2
C	10.5	7.6	4.8	7.9
D	9.6	6.2	0.8	7.4
Z	8.8	3.7	2.0	13.8

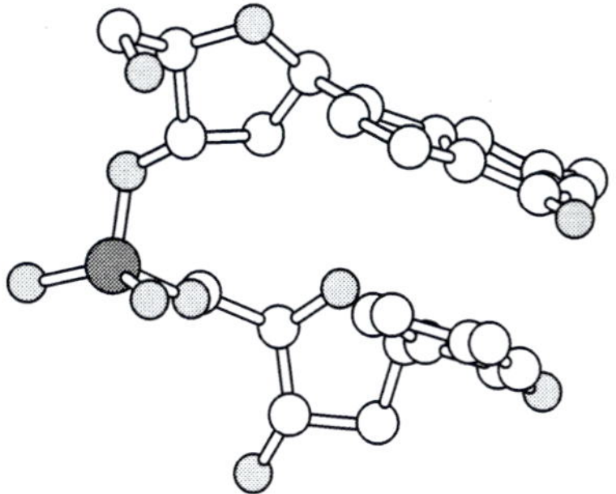

Fig. 3.4 Conformation of two successive nucleotides in one strand of canonical A-DNA, from fibre diffraction analysis.

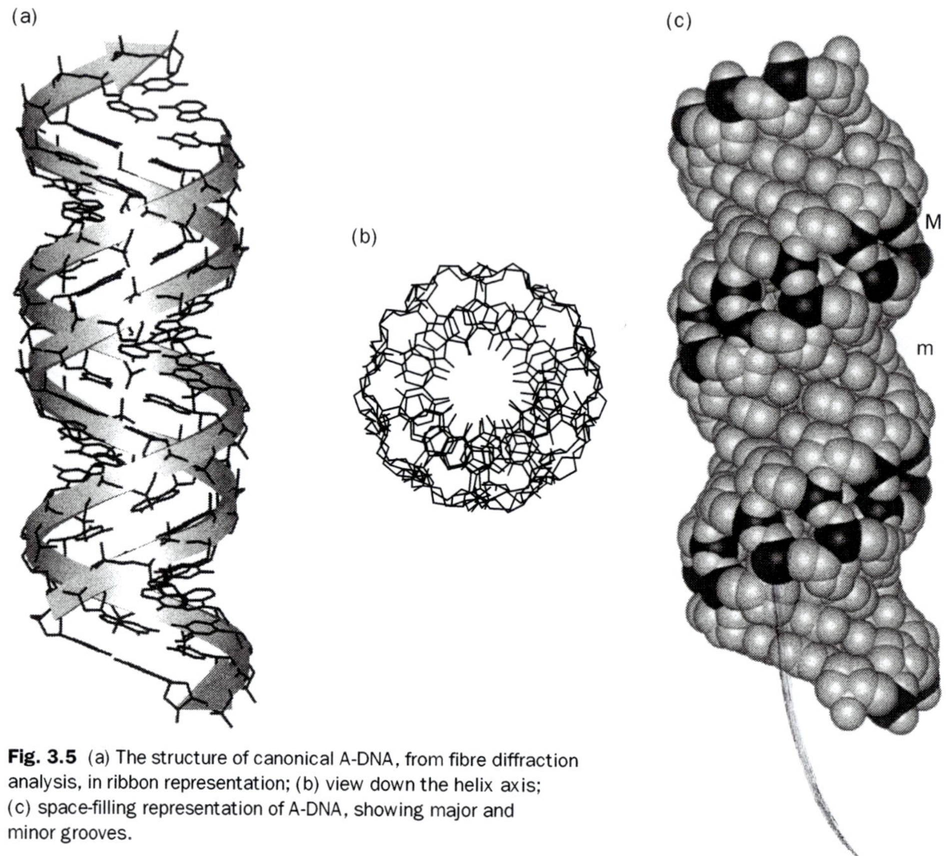

Fig. 3.5 (a) The structure of canonical A-DNA, from fibre diffraction analysis, in ribbon representation; (b) view down the helix axis; (c) space-filling representation of A-DNA, showing major and minor grooves.

twisted and tilted with respect to the helix axis (Fig. 3.5a), and are displaced nearly 5 Å from it, in striking contrast to the B helix. The helical rise is as a consequence much reduced, to 2.54 Å, compared to 3.4 Å for canonical B-DNA. The helix is wider than the B one and has an 11 base pair helical repeat. The combination of base pair tilt with respect to the helix axis and base pair displacement from the axis results in very different groove characteristics for the A double helix compared to the B form (Fig. 3.5b). This also results in the centre of the A double helix being a hollow cylinder (Fig. 3.5c). The major groove is now deep and narrow, and the minor one is wide and very shallow.

3.2.2 **DNA polymorphism in fibres**

The structural transition from A to B forms in 'mixed-sequence' DNA is induced by changes in the relative humidity surrounding the DNA fibres. Other forms can be produced by fine control of this condition (the C form) or from defined-sequence polynucleotides (the D and Z forms). For example, there are a number of members of both the A and B families that differ from the 'canonical' forms outlined above, in large part because of the degree of over-winding of the helices (for the B type family), and have considerable differences in pitch, base pair tilt with respect to the helix axis, and hence groove characteristics. There are variants on the A-DNA duplex that have a wide major groove,[4] reminiscent of that in a B helix, rather than that of the narrow standard A groove. Changes in groove width can thus be achieved at little energy cost.

The C form of DNA results from relatively low humidity conditions for a DNA fibre and is over-wound relative to B-DNA, with 9.3 residues per turn. The overall appearance of this helix and the dimensions of the grooves resembles B-type DNA rather than the A form morphologies. The D form cannot be adopted by random sequence native DNAs, and has been observed in crystalline fibres of the alternating purine/pyrimidine polynucleotides poly(dA–dT)•poly(dA–dT) and poly(dI–dT)•poly(dI–dT), where I is inosine (this is the rare 8-oxo-purine). The structure of this polymorph has not been unambiguously defined as yet, with models proposed ranging from left-handed seven- and eight-fold helices to a more conventional eight-fold right-handed one. Structural parameters for this right-handed model are given in Tables 3.1–3.3. Observations of the reversible structural transition from B to D forms in crystalline fibres have been made using time-resolved X-ray diffraction by means of a high-intensity synchrotron source of X-rays.[5] The observation of a gradual rather than an abrupt increase in helical pitch, from 24 to 34 Å, is consistent only with a transition in which there is no change in helix handedness, and so a left-handed D-DNA model can be rejected.

The Z form is an authenticated left-handed structure.[6] It exists in the alternating sequence poly(dC–dG)•poly(dC–dG) and is presumed to be the structure formed by this sequence in solution in high salt (>2.5 M NaCl) conditions. This left-handed DNA has a dinucleotide repeat (Fig. 3.6) with quite distinct nucleoside conformations for the guanosine compared to the cytosine residues. Each individual repeat has a helical rise of 7.25 Å, so that the rise between successive base pairs is half of this—3.63 Å (Fig. 3.7). Z-DNA is discussed further in Section 3.5; it was discovered in a defined-sequence oligonucleotide crystal structure as well as by fibre diffraction analysis of poly(dC–dG)•poly(dC–dG).

The polymorphism of DNA structure that is apparent in polymeric fibres, is suggestive of an underlying flexibility in DNA structure. This is a natural consequence of the large number of backbone and sugar conformational variables, which together with base pair flexibility, can result in double helices that are structurally distinct yet are equivalent in energetic terms. Fibre diffraction studies can only hint at the fine detail of this flexibility. So further significant

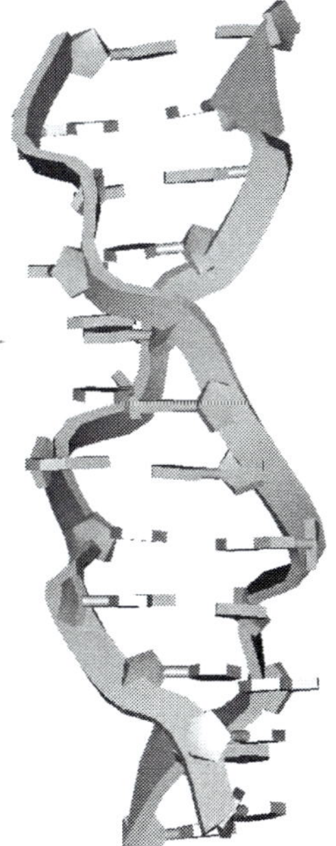

Fig. 3.7 The Z-DNA left-handed double helix, in ribbon representation to emphasise the discontinuities in the backbone.

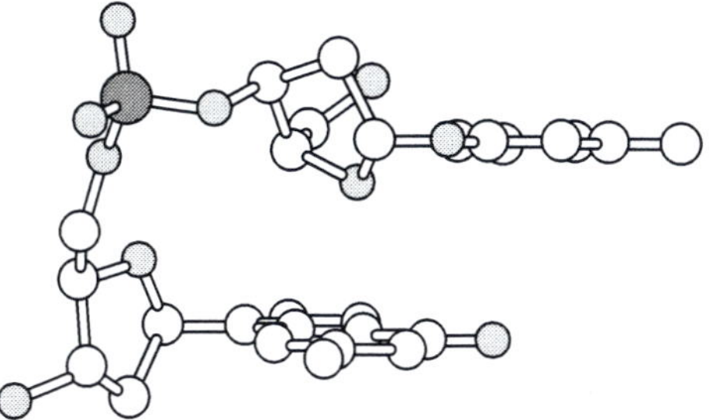

Fig. 3.6 Conformation of two successive nucleotides in one strand of canonical Z-DNA, from fibre diffraction analysis.

(a)

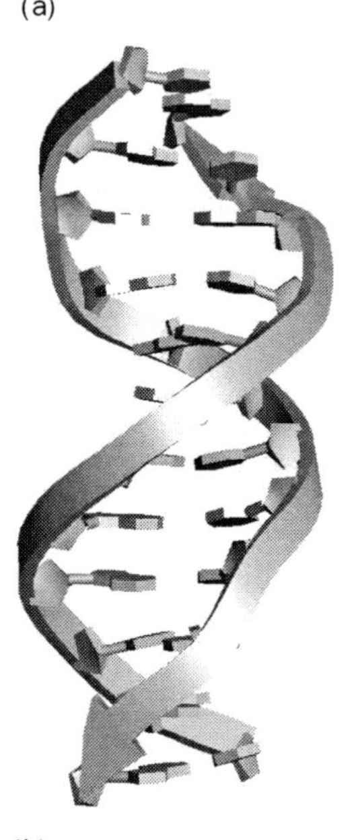

(b)

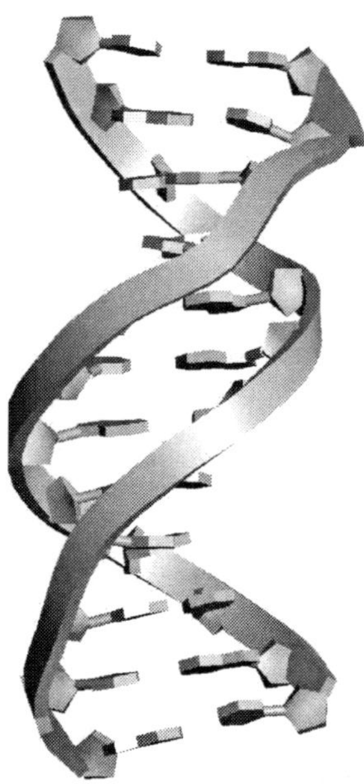

Fig. 3.8 (a,b) Two ribbon views of the structure of the Dickerson–Drew dodecamer, from the crystallographic analysis.[7,8] The two views are rotated by ~45°, and show the narrowing of the minor groove in the central region of the sequence as well as slight irregularities in the backbone conformation.

advances in DNA structure studies have taken place with the advent of single-crystal and NMR analyses of defined oligonucleotide sequences, which as described in Chapter 1, provide structural data that are not averaged over the sequence. These studies have provided much detail not only of the structures themselves, but also are enabling the principles to be determined that relate DNA structural features to sequence and environment.

3.3 B-DNA oligonucleotide structure as seen in crystallographic analyses

3.3.1 The Dickerson–Drew dodecamer

The single-crystal structure of the self-complementary dodecanucleotide d(CGCGAATTCGCG) was determined in 1979 by multiple isomorphous replacement methods.[7] This structure analysis is of historic importance in that without recourse to any preconceived model, it showed an anti-parallel right-handed B-DNA double helix (Fig. 3.8a), and thus unequivocally demonstrated the correctness of the Watson–Crick model for DNA structure. Thus, the 'Dickerson–Drew' sequence has subsequently been widely studied by a variety of other experimental and theoretical techniques, that have complemented, generally confirmed and sometimes extended the original structural results. The original crystallographic analysis, at 1.9 Å resolution, also revealed a number of major sequence-dependent structural features that could not be observed in fibre diffraction studies of averaged-sequence B-DNA, which has all the nucleotides in an identical conformation. These features are:

1. A narrow minor groove in the 5'-AATT region (Fig. 3.8b), which at its extreme point is only 3.2 Å wide compared to 6.0 Å in fibre diffraction averaged canonical B-DNA. The major groove at this point in the dodecamer structure is exceptionally wide (12.7 Å).

2. A well-ordered and regular network of water molecules in this A/T region of the minor groove.[8] This network has been termed the 'spine of hydration'. It involves hydrogen bonding of first-shell water molecules to the O2 atom of thymine and the N3 atom of adenine such that each water spans the two dodecamer strands. These water molecules are linked together by further, second-shell waters. Evidence for the persistence of

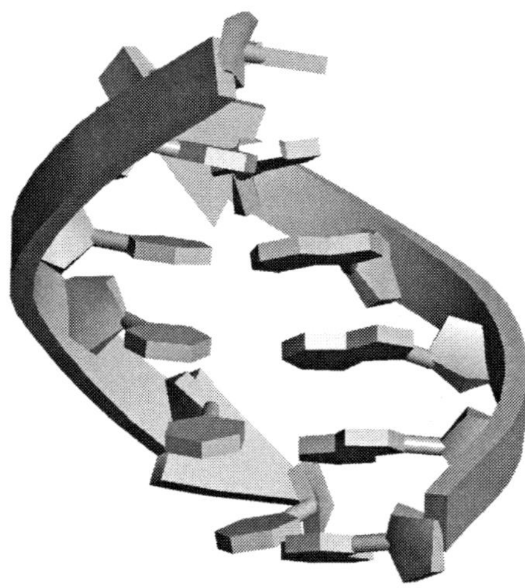

Fig. 3.9 View of the central region in the Dickerson–Drew dodecamer, showing the high propeller twist of the A•T base pairs.

this structured water arrangement in solution has come from NMR studies.[9] Minor-groove hydration and the structure of the spine are discussed in detail in Chapter 5.

3. Differences in the values for a number of local base pair and base-step parameters at different points along the sequence.[10] Most notable are the high propeller twist values for the four central A•T base pairs (Fig. 3.9).

4. A wide distribution of values for sugar puckers, glycosidic angles and backbone conformational angles. The value for a particular torsion angle is to a considerable degree dependent on the sequence context of the nucleotide involved. For example, values for torsion angles ϵ and ζ are distributed in two sub-groups within their broad ranges. Both angles are normally in the *trans, gauche⁻* domain, but are *gauche⁻, trans* when the nucleotide is followed by a purine. This alternative phosphate conformation has been designated B_{II}, with the standard *trans, gauche⁻* phosphate conformation being termed B_I. The B_{II} conformation is associated with increased minor groove width, both in the dodecamer and in a number of oligonucleotide crystal structures that have been subsequently determined. Several other correlations between backbone torsion angles were found in the original Dickerson–Drew structure, and subsequently confirmed in a number of oligonucleotide structures, such as those between angles α and γ, and between angle δ and glycosidic angle χ. The range of values for the δ angle is itself sugar-pucker dependent; this important correlation confirms and extends earlier findings from surveys of mononucleoside structures (Section 2.5), to the oligonucleotide level.

5. The *average* features of the structure are close to those for canonical B-DNA from fibre diffraction. For example, the average helical twist is 35.9° and there are 10.1 base pairs per helical turn.

6. The helix is not straight, but is bent by about 19° in the major-groove direction. This bending is apparent in Fig. 3.8b.

3.3.2 **High-resolution studies of the Dickerson–Drew dodecamer**

The role played by lattice interactions in the dodecanucleotide crystal structure has been examined in detail by several studies. Comparison of nine different dodecamers in three space groups has shown that the key parameters of minor-groove width and propeller twist for the central 5′-AATT region are largely unaffected by crystal packing factors.[11] This is to be expected since in the $P2_12_12_1$ original space group for the Dickerson–Drew (and a number of other) dodecamers, only the terminal two nucleotides at each end are involved in intermolecular interactions with other duplexes. Thus, the central helical region of 6–8 base pairs is unconstrained, and its features are intrinsic to the particular sequence. In a more demanding test of the generality of the features it has more recently been found that d(CGCGAATTCGCG) can crystallize in an alternative space group (the trigonal R3) with a distinct packing arrangement, when under the influence of particular metal ions such as calcium.[12,13] The key features of base and base pair morphology are still apparent, although sequence-dependent features tend to differ less from canonical values. These together with the narrow minor groove filled by a spine of hydration, provide good evidence for the general correctness of the original structure. All of these structures contain both dodeca-nucleotide strands in the crystallographic asymmetric unit, and the resulting duplex is not perfectly two-fold symmetric, by contrast with its structure in solution. A further recently discovered crystal form[14] in the trigonal space group $P3_212$, by contrast, has the duplex sitting on a crystallographic two-fold axis, so that exact two-fold symmetry is imposed on the structure (Fig. 3.10). Features such as the narrow minor groove and hydration spine are still apparent, but now they are two-fold symmetric, reflecting the exact symmetry implied by the self-complementary Dickerson–Drew sequence.

The status of the Dickerson–Drew structure as the paradigm for many of the fine details of B-DNA structure generally, implies that its features are considered to be totally reliable, especially by non-experts in crystallography. It is important to bear in mind that the original structure was determined to a resolution that is only moderate by current standards,

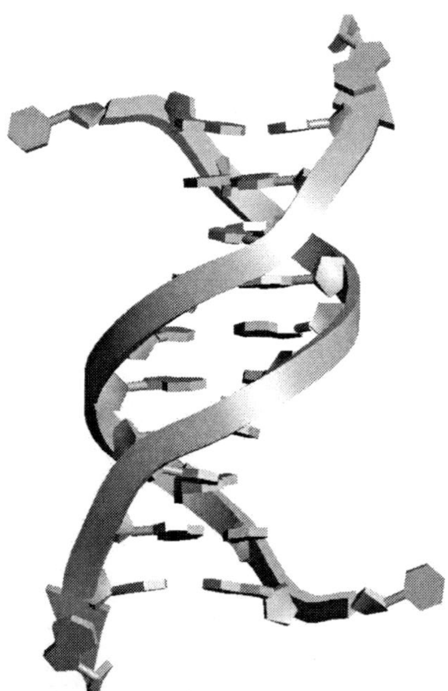

Fig. 3.10 The structure of the dodecamer d(CGCGAATTCGCG) in space group $P3_212$, where the duplex utilises a 2-fold crystallographic symmetry axis.[14] The terminal nucleotides are not base-paired within the duplex; instead they are hydrogen-bonded to other duplex molecules in the crystal lattice.

and refined by methods that are now considered to be sub-optimal. Thus, the accuracy and precision of this analysis are probably much lower than the more recent studies on this sequence. These have used (i) high quality, highly redundant diffraction data from synchrotron sources, (ii) libraries of high-precision nucleotide geometry, (iii) refinement methods that are able to overcome local false minima. The reassuring picture that is emerging is one in which the concept of sequence-dependent structure is retained.

The crystal structures of the sodium and magnesium forms of the Dickerson–Drew sequence have been determined to 1.4 and 1.1 Å resolution respectively.[13,15,16] This structure confirms many of the above details, apart from the assignment of solvent in the minor groove, with the analysis being consistent with primary solvent sites in the spine of hydration sharing between sodium cations and water molecules (the arrangement is described in detail in Chapter 5). Analysis of minor-groove width in the Dickerson–Drew structure using molecular dynamics simulations has led to the suggestion[17] that it changes as a result of cation interactions, and is thus dependent on the nature of the cation, as well as on sequence in the minor groove.

3.3.3 Sequence-dependent features

Typical variations in propeller twist and roll (see Chapter 2 for definitions) in dodeca-nucleotide B-DNA crystal structures, are shown in Fig. 3.11. The sequence-dependent effects are most reproducible for the central 6–8 residues of these sequences since the 3′ and 5′ termini of the dodecamer structures are involved in packing interactions with other molecules in the crystal. The pattern of base and base pair morphological param-eters is also consistent with those derived from NMR data in solution using the method of residual dipolar couplings.[18]

The most significant variations in values for these parameters from canonical values are:

1. The local helical twist varies by up to 15°, with pyrimidine-3′,5′-purine steps having lower than average values and purine-3′,5′-pyrimidine steps having higher than aver-age ones. These differences correlate with differences in the susceptibility of the sugar-phosphate backbone bonds in this sequence to be cleaved by the DNase I enzyme,[19] with the TpC step, which has a high twist, being the most easily cleaved. The mean twist angle in the crystal structure is 36°, in accord with the fibre diffraction value.
2. Propeller twists are significantly greater for A•T base pairs than for G•C ones, by an average of 5–7°.
3. Roll angles for pyrimidine-3′,5′-purine steps have positive values, since they open up toward the minor groove, whereas purine-3′,5′-pyrimidine steps have negative roll angles with major groove opening.

There is good evidence that at least some sequence-dependent features seen in the dodecamer structure have direct relevance to 'real' DNA sequences, and are not restricted to short crystalline ones, especially the larger scale features such as the groove width vari-ation. For example, it has been found[20] that there are variations in the ability of the DNase I enzyme to cleave the phosphodiester backbone of the 160 base pair *Escherichia coli tyr*T promotor within runs of A•T base pairs. These variations correlate with changes in minor-groove width that occur in them when TpA steps are present or absent.

The sequence-dependent changes in helical and propeller twist, roll and slide were initially rationalized on the basis of steric clashes between substituent atoms on individual

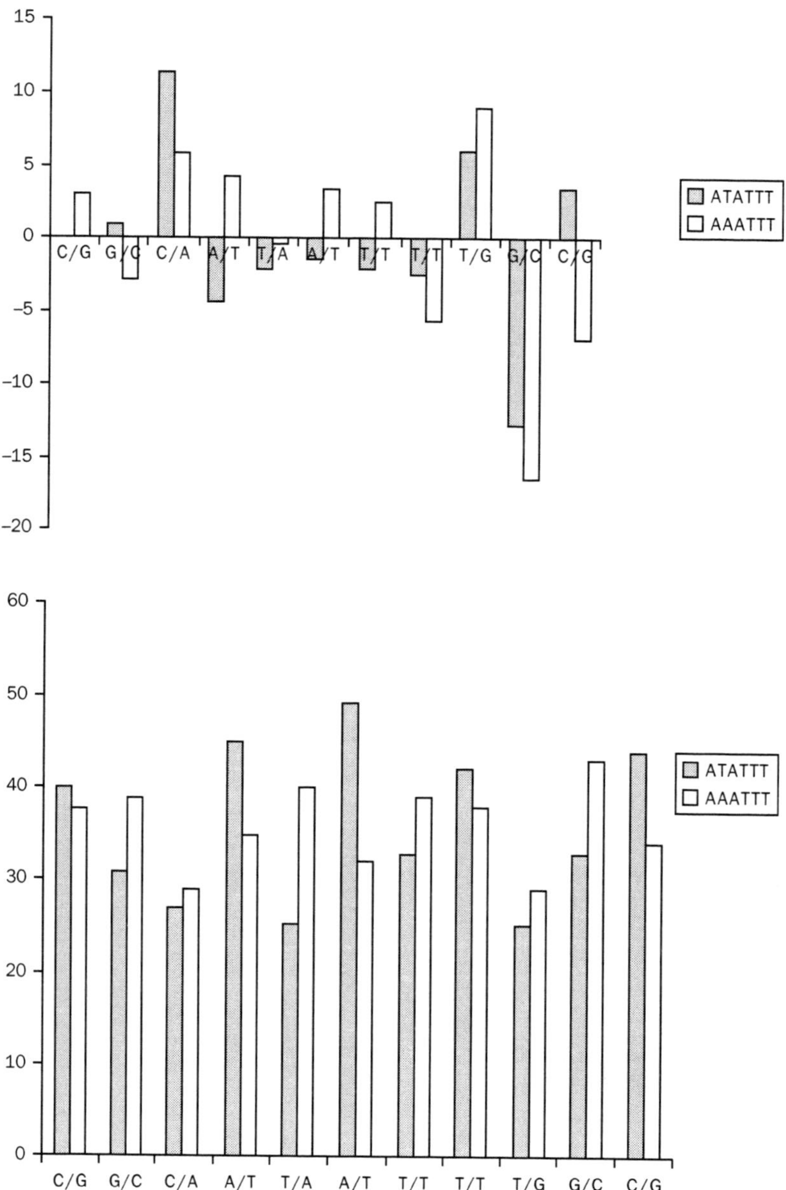

Fig. 3.11 Plots of the helical parameters propeller twist and roll in two dodecanucleotide crystal structures (see: Aymami, J., Nunn, C. M. and Neidle, S. (1999). *Nucleic Acids Research*, **27**, 2691).

bases—the Calladine Rules.[21,22] These rules are based on the premise that the observed structural effects derive from opposite-strand purine–purine steric hindrance, and thus give an essentially mechanical view of DNA structural features as arising from the interplay of rigid-body Newtonian forces. For purine-3′,5′-pyrimidine steps these clashes are between major-groove substituent atoms O6 of guanine and N6 of adenine, and between minor-groove N2 of guanine and N3 of adenine for pyrimidine-3′,5′-purine steps. On this basis, little steric clash would be predicted for purine-3′,5′-purine and pyrimidine-3′,5′-pyrimidine steps. The rules suggest that clashes are avoided by combinations of changes in twist, roll and slide. Their relative proportions depend on the nature of the bases involved:

- a decrease in local helical twist decreases minor-groove clashes
- a increase in roll angle in the groove with clashes
- separation apart of successive purines by means of an increase in slide between them
- a decrease in propeller twist. The changes in slide result in alterations in the backbone angle δ, which thus has a higher value for purine compared to pyrimidine nucleosides.

The Calladine Rules, which have reasonable predictive capability for some (though by no means all) B-DNA structures other than d(CGCGAATTCGCG), only have limited applicability to other helical types, especially A forms where there is inherently high propeller twist. The inability of the rules to take factors such as hydration, electrostatics and base stacking into account, does mean that they only provide a first-order approximation to the understanding of sequence-dependent effects, even in B-DNA structures.

There have been a number of subsequent studies that have examined relationships between parameters and their underlying structural basis, which have been able to use the increasing number of B-DNA structures available in the Database by searching for correlations between parameters.[23,24] Although the available structures do not enable all possible flanking sequences to be examined (see Section 3.3.4), it does appear that they only influence some individual base steps, notably pyrimidine-3′,5′-purine ones.[24] The sequence dependence of helical twist angles has been examined in 38 structures,[25] which has demonstrated a significant correlation between twist and roll, as a consequence of steric clashes, some of which are in addition to the purine–purine clashes defined by Calladine.[21] These authors have introduced an empirical clash strength function, which while able to reliably predict the twist of most steps, fails with some CpA ones.

A consideration of base–base interactions (rather than solely considering substituent interactions) by means of empirical energy functions has been a fruitful approach.[26,27] These functions include terms for van der Waals and electrostatic interactions, with charge distributions in bases being calculated with on the basis of a π-electron model rather than the more commonly used point charge one. Experimental values for rise, roll, and tilt, which are especially dependent on base stacking, are well reproduced by this approach. An extension of this approach has used a genetic algorithm to find global minima[28] for a set of 30 diverse oligonucleotide structures, all of which also have known crystal structures. Most of them are well-predicted, with the sequence-dependent trends in base-step and pair morphologies mostly being well-reproduced. However, since the accuracy of the structures themselves is unknown, correlations with computational predictions of the details of sequence dependency may be premature until sufficiently large samples of structures are used.

Backbone conformational angles show considerable variability in B-DNA structures with respect to sequence (Table 3.4). Even within the Dickerson–Drew structure, individual angles typically show a $>45°$ spread of values along the sequence. The central question, that is common to variations in backbone angles and base morphological parameters is whether these differences reflect (i) true sequence-dependent variability,

Table 3.4 Ranges of backbone torsion angles (°) observed in high-resolution (better than or equal to 1.9 Å) B-DNA crystal structures, from ref. (29). The ranges corresponding to B_I and B_{II} phosphate conformations are indicated

	α	β	γ	δ	ϵ	ζ	χ
Range	270–330	130–200	20–80	70–180	160–270	150–210 230–300	200–300
Mean values	298	176; B_I 146; B_{II}	48	128; B_I 144; B_{II}	184; B_I 246; B_{II}	265; B_I 174; B_{II}	258; B_I, Pu 241; B_I, Py 271; B_{II}, Py

(ii) the influence of crystal packing interactions, or (iii) are a reflection of a lack of precision in the crystal structure determinations. Surveys of conformational angles[29] from a large number of crystal structures show that each angle occupies a discrete range (Table 3.4), with some such as ζ occupying two regions. The angles have Gaussian distributions, suggesting that the variations reflect both experimental inaccuracies and real fluctuations.

The experimental data does not point to clear sequence dependency relationships between backbone conformation and helical parameters, in broad agreement with a theoretical analysis,[30] which concludes that roll, tilt and rise are inherently backbone-independent. Helical twist is dependent on backbone conformation, though not in a straightforward way.

3.3.4 **Other B-DNA oligonucleotide structures**

The Calladine rules provide a means of understanding the structural behaviour of dinucleotide base steps, of which there are just 10 distinct types. The rules imply that each step is largely independent of its neighbours, a considerable over-simplification. Extension to four-base sequences would require information for 136 distinct possible combinations. The establishment of the first B-DNA oligonucleotide structure provided information on only a very small number of these possible nearest neighbour combinations of DNA sequence, and so there have been concerted attempts to determine structures for other sequences, both relating to and distinct from the Dickerson–Drew one. Even though several hundred B-DNA crystal structures have now been determined (at the time of writing there are 230 in the NDB), less than half of the 136 tetranucleotide combinations have been sampled experimentally. A recent theoretical study of sequence-context effects at the tetranucleotide level[31] used an experimental database with 66 out of the 136 included. This study suggests that A/T-containing dinucleotide steps tend to be significantly more context-independent than C/G-containing ones.

The crystal structures of standard B-type oligonucleotides have mostly averaged features that closely approximate those of averaged-sequence fibre-diffraction B-DNA. None is more than 12 base pairs in length. Almost all are self-complementary palindromic sequences. They fall into two principal categories. They are either dodecamers, the overwhelming majority of which are isomorphous to the Dickerson–Drew structure and therefore pack identically in the crystal, in the space group $P2_12_12_1$, or they are decamers, which crystallize in a variety of space groups, some of which give very high-resolution

Table 3.5 Selected B-DNA crystal structures

Sequence	NDB number
d(CGCGAATTCGCG), orthorhombic	BLD0001
d(CGCGAATTCGCG), trigonal	BD0032
d(CGCAAAAAAGCG)	BDL006
d(CGCAAAAATGCG)	BDL015
d(CCGCTAGCGG)	BD0028
d(CCAACGTTGG)	BDJ019
d(CCAGTACTGG)	BD0023
d(ACCGGCGCCACA)	BDL035
d(CATGGGCCCATG)	BD0026

(0.74–1.4 Å) structures (Table 3.5). The isomorphism of the dodecamers provides a convenient framework to analyse systematic changes in the central base pairs, while keeping invariant the 3′ and 5′ terminal sequences, which are involved in the crystal-packing arrangements. Several decamers crystallize in forms with the 10-mer duplexes packed end-to-end in the crystal, in a quasi-helical manner.

A number of changes to the central sequence of the Dickerson–Drew dodecamer have been examined. When the sequence is a purely alternating one, as in d(CGCATATATGCG),[32] the minor-groove width is equally narrow as in the Dickerson–Drew structure, although the narrowest points do not coincide. This 5′-ATATAT sequence shows a pronounced alternation of several of its base pair and step parameters, especially roll; propeller twist values for the A•T base pairs are consistently lower than those in the Dickerson–Drew sequence.

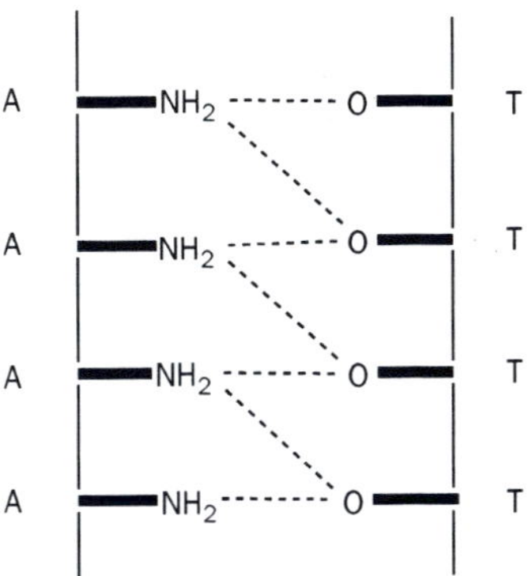

Fig. 3.12 Schematic view of three-centre major groove bifurcated hydrogen bonding involving A•T base pairs.

The structure of the duplex formed by d(CGCAAAAAAGCG) and its complementary sequence[33] is remarkable for several reasons. It is a model sequence for an A tract, which in natural DNA has anomalous structural and dynamic properties—it cannot readily be reconstituted into nucleosomes, and is intrinsically bent when in phase with a helical repeat (see Section 3.6). The structure has high propeller twists for the A•T base pairs, of up to 26°, with a network of bifurcated three-centre hydrogen bonds connecting them (Fig. 3.12). These bifurcated hydrogen bonds are in the major groove and involve atoms N6 of adenines and O4 of thymines. They have been suggested to be major factors in stabilizing the high propeller twists for the A•T base pairs, and thus in stiffening the A tract, at least in this structure. The roll, tilt and slide values for the ApT steps in the structure are all close to 0°, with bending (towards the major groove) occurring at the ends of the A tract rather than within it. Bending in the major groove direction has also been observed in the structure of the duplex formed by d(CGCAAAAATGCG) and its complementary strand,[34] although the situation here is complicated by the presence of two distinct orientations of the duplex in the crystal lattice. This study has concluded that the bending found in the isomorphous dodecamer crystal structures, is more a consequence of crystal packing forces than of intrinsic properties of the A tract sequences. The high (~20°) propeller twists for A•T base pairs found in the structure of d(CGCAAATTTGCG), are not accompanied by significant three-centre hydrogen bonding in the A/T region, suggesting[35] that these effects may not be inherently required in all structures to stabilize high propeller twist and a narrow minor groove. Rather, high propeller twist is more a consequence of base stacking preferences.

The weight of evidence supports the concept of a narrow minor groove being preferred in A/T regions. However this is only a preference, and not a definitive rule obeyed under all circumstances. On the one hand, the duplex formed by d(CGCAAGCTGGCG) has an average minor-groove width of about 7 Å in the 5′-AGCT region.[36] This is close to that of averaged fibre diffraction B-DNA, as is the G/C-rich minor groove in the d(ACCGGCGCCACA) duplex.[37] On the other hand, the high-resolution (1 Å) structures of the decamers d(CCAACGTTGG) and d(CCAGCGCTGG) both show a pronounced narrowing in the central region of the minor groove,[38] even though individual base pair propeller twists are generally no greater than about 12°. The flexibility of such structures is further shown in that[39] of the decamer d(CCAGTACTGG). This is notable in being the sole DNA duplex crystal structure to be determined at true atomic resolution (0.74 Å), so that the accuracy and precision of atomic coordinates and derived parameters is fully established. (The average esds of bond lengths and angles are 0.028 Å and 2.19° respectively). The structure shows that a number of regions of the backbone adopt more than one conformation, an effect that is undetected at lower resolution. These alternative conformations do not affect the general pattern of minor-groove narrowing that is centred around the short two base pair A tract. These A•T base pairs have high propeller twists of −20.3° and −18.8°, whereas the two adjacent C•G ones have values of −8.4° and −9.7° respectively.

The non-palindromic sequence d(ACCGGCGCCACA)[37] is a hot-spot for frameshift mutagenesis by several chemical carcinogens. The structure is non-isomorphous with the other dodecamers and is remarkable for several features. Three G•C base pairs in the centre of the helix are partially opened, with exceptionally high propeller twists and some loss of G•C hydrogen bonding. It has been suggested that this represents the intermediate stage prior to full base pair opening. Analysis of a low temperature form of the crystals has shown[40] that cooling produces a one base pair shift of the major groove base pairing that is favoured by CA tracts. The crystal packing involves helices interacting together by means of groove–backbone interactions,[41] which can be considered to be a precursor model for true DNA–DNA junctions such as the Holliday junction involved in recombination processes (see Chapter 4).

Crystallographic analyses of a number of the diverse sequences that are able to crystallize as decamer duplex helices (refs 42–45) has provided further evidence for the variations in sequence-dependent structural features. Comparisons of the structures show that the parameters helical twist, roll, and rise are not independent of each other, but that there are several correlations between them. For example, rise is linearly related to helical twist. The TpA step, even though it is consistently found to be unstacked in all the B-DNA oligonucleotide structures where it occurs, has highly variable sequence and crystal environment-dependent behaviour. An extreme and surprising example of this is in the sequence d(CGATATATCG), which has a wide minor groove in the alternating AT region,[46] albeit with a characteristic alternation of helical twist angles that is a consequence of the alternation of good base stacking at ApT steps combined with the poor stacking at the TpA ones. Thus, although alternating A/T sequences generally have a marked tendency to produce narrow minor grooves in B-DNA, their inherent structural flexibility means that this tendency can be readily overcome under the influence of TpA unstacking. The B_{II} phosphate conformation, which occurs only at steps with the second base being a purine, is itself a factor in producing a widened minor groove.[47] However the B_I/B_{II} backbone conformational flexibility observed in several high-resolution crystal structures[39,48] suggests that this is less a factor than the inherent properties of the bases and base steps themselves.

3.4 **A-DNA oligonucleotide crystal structures**

DNA double helices of the A type are produced in fibres of random-sequence DNA under conditions of low humidity (see Section 3.2), and in solution when the water activity is reduced by the addition of various alcohols. The fibre diffraction studies suggest that certain runs of sequence such as alternating G•C base pairs, may have a tendency to be in an A form. Single-crystal analyses of particular lengths of oligonucleotide sequences have determined many to be of A type, especially when they have high G/C content. However, for the majority of such structures, it is now apparent that this does not reflect so much any inherent structural preference, as the crystal packing requirements for these particular lengths of oligonucleotide duplexes. The high concentrations of organic alcohols commonly used in oligonucleotide crystallizations may also be a contributory factor. Nonetheless, these structures have provided invaluable insights into the flexibility and conformational preferences of A-DNA helices in general, even if their significance in biological terms is less clear. There is at least one instance where A-type structures are of undoubted importance: this is in the case of RNA–DNA hybrids, formed, for example, during transcription and replication. Since RNA helices are always constrained to be in an A form, then DNA–RNA hybrids would be expected to be similarly constrained. This has been found to be the case by both fibre diffraction studies of RNA–DNA hybrid polynucleotides and by single-crystal analyses. For example, the structure of the duplex formed by r(GCG)d(TATACGC) shows that the DNA strand has adopted a conformation close to that of duplex RNA, with an 11-fold helix (average helical twist of 33°), C3′-*endo* sugar puckers and an A-type helical backbone conformation.[49]

3.4.1 **A-form octanucleotides**

A large number of self-complementary octanucleotides, of widely varying sequence type, have been crystallized and found to have A-DNA family structures (Table 3.6). As with B-type oligomers, structural features averaged over a number of such sequences are close to those of fibre diffraction A-DNA. By contrast with B-DNA oligomer crystal structures, backbone torsion angles tend to have narrow ranges of values, reflecting the greater conformational rigidity of the A form. Some of the A-form octamer sequences crystallize in the tetragonal space group $P4_32_12$; most of the others are in the hexagonal one $P6_1$. It has been possible to vary conditions such that some sequences such as d(GGGCGCCC)[50] and d(GTGTACAC)[51] have been crystallized in both forms. The hexagonal form has

Table 3.6 Selected A-DNA crystal structures

Sequence	NDB number
d(GGGCGCCC), tetragonal	ADH026
d(GGGCGCCC), hexagonal	ADH027
d(CCCCGGGG)	ADH056
d(ACGTACGT)	ADH070
d(CCGGGCCCGG)	ADJ082
d(GCGTACGTACGC)	ADL046
d(CGCCCGCGGGCG)	ADL047
d(CATGGGCCCATG)	BD0026

also been found for sequences such as d(GGGGCCCC),[52] d(GGGTACCC)[53] and d(GGGATCCC)[54] as well as for a number of mismatch variants. All have helical and conformational features within the general A class, but with some marked local variations, in particular in the overall helical repeat, in minor-groove widths and in base step parameters. The wide minor groove, of ~15 Å in d(GGGGCCCC) contrasts with the more typical value of 9.7–9.8 Å in d(GGGTACCC), which is close to that for canonical fibre diffraction A-DNA, of 11.1 Å. Major-groove width cannot be fully assessed in an octanucleotide duplex, since eight rather than seven phosphate groups are required in order to calculate the shortest phosphate–phosphate interstrand distance. Approximated major-groove widths show very wide variation, ranging from 5 Å in the low-temperature form of d(GGGCGCCC)[55] to over 12 Å in d(GGGGCCCC),[56] compared to the very narrow fibre value of 2.2 Å. The tetragonal form, with sequences having a pyrimidine-3′,5′-purine step at positions 4 and 5, that is at the centre of the helix, shows significant deviations from standard A-form backbone geometry at this point, with torsion angles α and γ having *transoid* values rather than the normal *gauche⁻* and *gauche⁺* ones respectively. This discontinuity has been observed in several structures, for example that of d(ATGCGCAT).[55] It is likely that this effect is a consequence of the requirements imposed by hydration and crystal packing in the tetragonal unit cell, rather than being an intrinsic property of these sequences.

3.4.2 **A-form oligonucleotides in solution? Crystal packing effects**

Indeed, crystal packing factors may well be the driving force behind *all* octanucleotide structures being in the A form. Solution NMR studies on the sequence d(ATGCGCAT)[55] have demonstrated that it shows B-family behaviour in solution even though the crystal structure is unequivocally of the A type. This dichotomy between crystal and solution environmental constraints is also shown by the sequence d(GGATGGGAG), which forms part of the binding site for the transcription factor TFIIIA that forms part of the initiation complex for transcription of the 5S RNA gene in *Xenopus*. The (low-resolution) 3 Å crystal structure of the duplex formed by this sequence and its complementary strand, shows an A-family helix,[56] with average major- and minor-groove widths of 14.8 and 15 Å respectively, and a helical repeat of 11.5 base pairs. These values closely correspond to those for A'-RNA from fibre diffraction studies. Nuclease digestion studies, showing a repeat of 11.4 base pairs per turn, together with circular dichroism measurements,[57] are consistent with the assignment of a structure for the d(GGATGGGAG) sequence as well as for the complete 54 base pair TFIIIFA binding site, that is not classical B-DNA. On the other hand, NMR and other biophysical methods all indicate that the nonamer in solution has normal C2'-*endo* sugar puckers and B-DNA range glycosidic angles.[58]

Short A-DNA double helices are insufficient in length for the major groove to be fully formed, and so estimates of its dimensions are necessarily approximate. X-ray analyses of two dodecamers (in two distinct space groups) have each revealed the structure of a complete turn of A-DNA double helix, as well as conformational features that are not subject to the crystal packing factors of the octanucleotides. The structure of the sequence d(CCCC-CGCGGGGG), with both alternating and non-alternating nucleotides[59] has average backbone conformational angles and helical parameters that are remarkably close to those in A-DNA fibres, with some local backbone angle variations, for example, *trans* α and γ angles at one of the two CpG steps (Fig. 3.13). This structure, together with

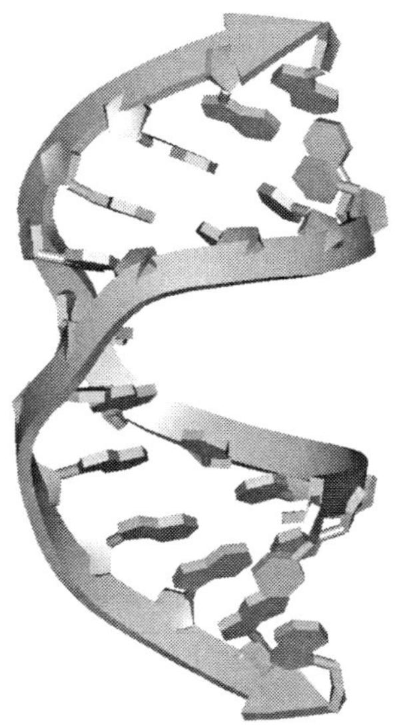

Fig. 3.13 Structure of a complete turn of A-form DNA, as found in d(CCCCCGCGGGGG).[59]

others such as that of the A-DNA decamer d(ACCGGCCGGT),[60] is consistent with the notion that very GC-rich sequences have a higher tendency to form A-type duplexes. The sequence d(CCGTACGTACGG), of which two-thirds consists of G•C base pairs, likewise has an A-DNA structure,[61] this time without any *transoid* α/γ angles. The average major-groove width in this structure, of 3.5 Å, is similar to the 2.2 Å value from fibre diffraction. A high-resolution crystallographic study of the sequence d(AGGGGCCCCT) in two distinct space groups concludes[62] that conformation is less conserved in this circumstance than when different A-type sequences crystallize in the same space group.

Changes in crystallization conditions do not necessarily result in changes to the resulting helical structure. This has been demonstrated in a study[63] of the structure of the sequence d(GACCGCGGTC), crystallized under two contrasting conditions of ionic strength. This sequence, which is half of the human papilloma virus E2 binding site, forms an A-type helix with closely similar helical parameters under both conditions. However there is a marked difference in the degree of helical bending, with that grown under higher salt conditions being bent by 31° in the major-groove direction compared to 15° for the low-salt form. Such bending may be relevant to the need for flexibility when particular sequences bind to their cognate proteins, as we shall see in Chapter 7. Hydration in A-DNA oligonucleotides is generally extensive, with observations of phosphates and bases hydrogen-bonding to water molecules.[64] Most solvent ordering occurs in the major groove, with polygons of water molecules being observed in some crystal structures.[65] These polygon arrangements are especially apparent at high resolution.[66]

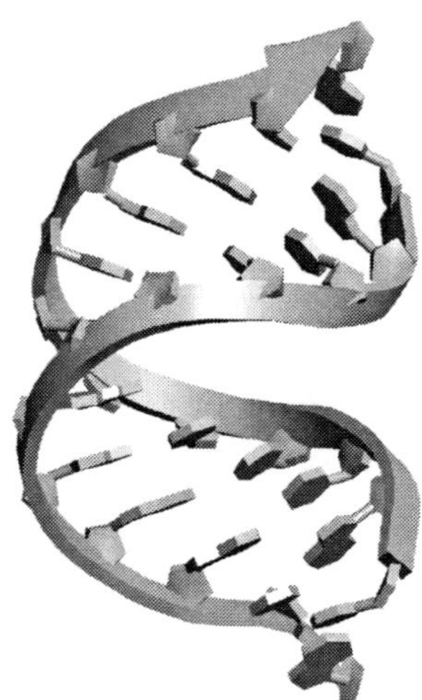

Fig. 3.14 Structure of the A/B hybrid d(CGCCCGCGGGCG), with the large bend at the centre of the structure being apparent.[68]

3.4.3 The A↔B transition in crystals

There are now almost 70 A-DNA crystal structures in the NDB, which provide a very detailed view of this DNA form and the myriad of deviations from canonical ideality (notably in major-groove width) that are possible.[67] It may be thought that there is then little more to learn about A-DNA. However the capacity of DNA structure to surprise is still apparent. The crystal structure of the dodecamer d(CGCCCGCGGGGCG) is remarkable in that parts of the structure demonstrate true hybrid character.[68] The central octamer region has an A-DNA conformation, but with a high degree of bending around the central purine-3′,5′-pyrimidine base step (Fig. 3.14). This 65° bending results in very close phosphate groups. By contrast the terminal nucleotides have many B-DNA characteristics, such as B-like sugar puckers, backbone conformational angles and base slide. This structure thus has features akin to an A↔B transition, albeit with distinct conformations at different points along the sequence. Whether this structure is a consequence of the influence of crystal packing forces or innate tendencies of this sequence is not clear.

The high-resolution structure of d(CATGGGCCCATG) has features that are throughout its length intermediate between A and B,[69] and thus resembles a trapped intermediate in the A↔B transition. There is a significant base shift, producing an A-like channel in the centre of the helix, and groove widths are intermediate between the A and B ideal values. A complementary view is provided by the analysis of six d(GGCGCC) duplex hexanucleotide crystal structures with varying cytosine brominations or methylations.[70,71] This results in differing crystal forms and structures that progressively show features that span A, B and plausible intermediate states. For example, the major groove becomes progressively deeper and values for the slide parameter become increasingly negative on proceeding from the B-type to A-type structures.

3.5 Z-DNA—left-handed DNA

3.5.1 The Z-DNA hexanucleotide crystal structure

The first oligonucleotide duplex single-crystal structure to be solved (by multiple isomorphous replacement methods) was that of the alternating pyrimidine–purine sequence d(CGCGCG) in 1979, at the very high resolution of 0.9 Å.[72] This structure is remarkable in that even though it consists of a duplex formed by the two anti-parallel hexamer strands, the helix is (quite unexpectedly) a left-handed one (Fig. 3.15). The backbone is irregular compared to A or B-DNA since the dC and dG residues have very distinct conformations (Fig. 3.6), resulting in a 'zig-zag' arrangement of phosphate groups—hence the helix has been termed Z-DNA. The same left-handed arrangement has also been found[6] in fibres of the alternating polynucleotide poly(dC–dG)•poly(dC–dG)—see

Section 3.2.2. Z-DNA was in effect discovered some years earlier during the course of circular dichroism studies on poly(dG–dC)•poly(dG–dC), when it was observed[73] that on increasing salt concentration beyond ~4 M NaCl, the CD spectrum became inverted from its standard B-DNA-associated shape. This major spectral change indicates a transition to a quite different conformational form, which is now accepted to be Z-DNA. A number of other physico-chemical techniques have subsequently been used to study this left-handed structure in addition to X-ray techniques.[74]

3.5.2 Structural features

Z-DNA oligonucleotides (and the poly(dG–dC)•poly-(dG–dC) polynucleotide) have the following conformational characteristics:

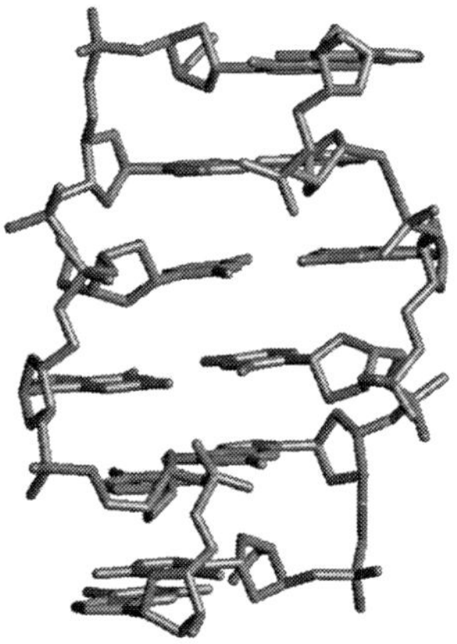

Fig. 3.15 Structure of the Z-DNA sequence d(CGCGCG).[72]

1. The purine (deoxyguanosine) nucleosides have *syn* glycosidic angles with a χ range of 55°–80° (mean 60°), together with C3′-*endo* sugar puckers. The conformation about the α/γ backbone torsion angles is *gauche*$^+$, *trans*, and *gauche*$^-$ about angle ζ.
2. The pyrimidine (deoxycytidine) nucleosides have *anti* glycosidic angles with a χ range of $-145°$ to $-160°$ (mean $-152°$), and C2′-*endo* sugar puckers. The α/γ conformation is *trans*, *gauche*$^+$ and that about angle ζ is *gauche*$^+$.
3. The G•C base pairs are of standard Watson–Crick type.

A consequence of these differences between purine and pyrimidine nucleosides is that the helical repeating unit is forced to be the CpG *di*nucleoside, rather than the *mono*nucleoside one in standard right-handed canonical A- and B-DNA. The distinct α/ζ phosphate conformations of the two nucleosides result in the characteristic zig-zag appearance of the backbone. Detailed examination of d(CGCGCG) in various crystal forms[75] has shown that there is a secondary backbone conformational family with a distinct set of phosphate orientations. This secondary conformation is termed Z_{II}, with the standard type as described above being termed Z_I. The Z_{II} conformation results in purines having a *gauche*$^+$ rather than a *gauche*$^-$ value for the ζ angle and a *gauche*$^-$ rather than a *trans* value for angle ϵ. Z_{II} pyrimidines have (rather smaller) changes in angles α and β compared to Z_I ones. Occurrence of the Z_{II} conformation at a particular residue is probably related to the coordination of the phosphate group at this point to a hydrated magnesium ion[75,76]—Z-DNA oligonucleotides are usually crystallized in the presence of magnesium and spermine ions (Table 3.7). Thus, the Z_{II} conformation does not occur at this point in the sequence in the pure-spermine, magnesium-free structure of d(CGCGCG).

3.5.3 The Z-DNA helix

The Z-DNA double helix is to a large degree represented by the structure of the hexanucleotide duplex d(CGCGCG); however there are slight differences between individual

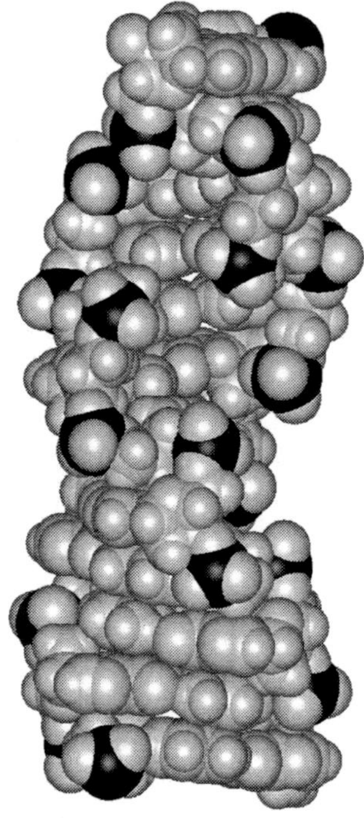

Fig. 3.16 van der Waals surface representation of a Z-DNA left-handed duplex, with phosphorus atoms highlighted in black.

Table 3.7 Selected Z-DNA crystal structures

Sequence	NDB number
d(CGCGCG)	ZD0003
d(CGCGCG), +Mg2+, spermine	ZDF001
d(GCGCGCGCGC)	ZDJ050
d(CGCGTG)	ZDF013

CpG units, quite apart from the occurrence of Z_I and Z_{II} forms. Idealized helices for both of these have been generated.[75] The fibre diffraction model for Z-DNA is closest to the Z_I helix. This Z helix has a 44.6 Å pitch with 12 base pairs per helical turn and a diameter of ~18 Å (Fig. 3.7), making it somewhat slimmer than a B-DNA helix (with a diameter of ~20 Å). The helical twists for two base pair steps CpG and GpC are quite different as a consequence of the asymmetry in guanosine and cytidine conformations, with the CpG step having an exceptionally small twist and almost no base stacking between the two C•G base pairs in the step. A further consequence of the *syn* guanosine conformation is that the base pairs are not positioned astride the helix axis, as in B-DNA. Rather, their edges are at the surface of the helix. The N7 and C8 atoms of the imidazole 5-membered ring of guanine and (to a lesser extent) the C5 atom of cytosine, actually protrude onto the helical surface of what would be the major groove, making the surface slightly convex at this point (Fig. 3.16). In other words, Z-DNA does not have a major groove at all. Its minor groove is very narrow and deep, and lined with phosphate groups (Table 3.3). The differences in phosphate orientation between Z_I and Z_{II} helices result in the latter having a ~1 Å wider minor groove.

3.5.4 **Other Z-DNA structures**

Z-DNA oligonucleotides have been crystallized with a range of related sequences, including pure analogues such as d(CGCGCGCG)[77] and d(CGCGCGCGCG),[78] as well as variants where changes have been made to base and sequence type. All show the resilience of the Z-DNA structural entity. The central CpG can be replaced by TpA while maintaining the left-handed structure.[79] This shows that Z-DNA can tolerate some A•T base pairs, although they do tend to destabilize the structure. The cytosines are required to be 5-methylated in order for stabilization of the d(CGTACG) sequence, and in general modified cytosines are necessary if a Z-DNA structure contains A•T base pairs. It has been suggested[79] that A•T base pairs are not able to take part in the ordered Z-DNA groove hydration that plays an important role in maintaining the integrity of the Z-DNA structure,[76] by contrast with G•C base pairs. Hence, the former cannot by themselves form a Z-DNA structure. The slight preference of guanosine nucleosides (unlike adenosine) to adopt the *syn* rather than the *anti* glycosidic conformation (see Chapter 2) is also a factor in the GC preference of Z-DNA. It is then surprising that reversal of the central pyrimidine-3′,5′-purine sequence (i.e. so that it is no longer purely alternating) in the structure of d(CGATCG), still retains a Z-DNA type structure,[80] albeit with the thymidines and adenines adopting *syn* and *anti* glycosidic conformations respectively. This energetically unfavourable Z structure was only stabilized by having the cytosines methylated or brominated at the 5 position of the base. Nonetheless, its existence suggest that the requirement for the formation of a Z-DNA structure is not so much that the sequence should be an alternating C/G-rich one, but that it retains the key feature of alternating *syn* and *anti* nucleosides. It is likely that replacement of thymine by uracil would increase Z-form stability, since the methyl group of the former presents a significant hydrophobic hindrance to solvent ordering around the Z helix. This has been borne out by subsequent structural studies on uracil-containing Z sequences.[81] In general non-alternating sequences can form Z-DNA structures in the crystalline state when the non-alternating pyrimidines are able to adopt a *syn* conformation.[82]

3.5.5 Biological aspects of Z-DNA

The accessibility of the C8 (and to a lesser extent the N7) position of guanine in Z-DNA renders it susceptible to attack by several electrophilic compounds that have a preference for these positions. Examples include the N-aryl carcinogens acetylaminofluorene and aflotoxin, as well as C8-bromination. All stabilize Z-DNA and promote the B to Z structural transition. Thus, Z-DNA tracts within genomic sequences could act as mutational hot-spots for these agents. The fact that Z-DNA formation is also facilitated by methylation at the C5 position of cytosine, a possible mechanism in some eukaryotic organisms for regulating gene transcription, suggests that Z-DNA could play a role in gene regulation.

Z-DNA in its linear form is less stable than either A- or B-forms, requiring high salt or alcohol concentrations for maintenance of its structure in solution. These reduce electro-static interactions between interstrand phosphate groups, which in Z-DNA are much closer together than in B-DNA (7.7 Å in the Z_I form compared to 11.7 Å across the minor groove in B-DNA). Z-DNA sequences can also be stabilized at physiological salt levels when incorporated in underwound, negatively supercoiled covalently closed-circular DNA, for example, in plasmid DNA. Even low (~1–2 per cent) levels of Z-forming sequence can significantly change supercoiling properties. It has been proposed[83] that a role for Z-DNA *in vivo* is to act as a signal for the induction of transcription via supercoiling.

The classic Z-forming sequence d(CG)$_n$ is much rarer in eukaryotic than in prokaryotic organisms. However, the sequence d(CA/GT)$_n$, which can also form left-handed DNA in negatively supercoiled plasmids, is much more prevalent in eukaryotic genomes. The overriding question of whether Z-DNA *actually* exists naturally, rather than that it *can* be formed under appropriate circumstances, has not as yet been fully answered. A number of proteins that bind to Z-DNA have been isolated, although their role(s) and significance have been unclear until recently. The crystal structure of a Z-DNA complex with the Zα high-affinity binding domain of an RNA editing enzyme has been reported.[84] A subsequent database search for other proteins with the same Z-binding motif found a closely similar domain in a tumour-associated protein (DLM-1); the crystal structure[85] of a Z-DNA complex with this shows close structural analogy with the Zα complex. In both the critical protein to Z-DNA contacts are made by three conserved amino acid residues. In both cases the nature of the protein suggests that Z-DNA plays a role in regulation of transcription/translation.

3.6 **Bent DNA**

3.6.1 **DNA periodicity in solution**

DNA under physiological solution conditions is generally assumed to be in a B-type conformation, even when combined with chromosomal proteins. There are at first sight some important differences between the B structures as found in the crystal and fibre, and that in solution, even as naked DNA. The structural studies all point to a helical repeat of 10.0–10.1 residues per turn for averaged-sequence DNA. However, plasmid mobility, nuclease digestion and hydroxyl radical cleavage experiments[86] have given a rather different value, of close to 10.5 base pairs per turn, again averaged over many base pairs. This apparent discrepancy is only partly resolved by the observation that helical periodicity in solution is markedly sequence dependent.[87] Cleavage at a particular point along a DNA sequence depends on the local helical twist and hence groove width, in a manner that correlates with the features that have been found in the Dickerson–Drew dodecamer, both in the crystal and in solution by NMR.

3.6.2 **A tracts and bending**

The DNA double helix is usually thought of as a straight molecule. The original Watson–Crick model is of a straight helix, since it was based on data from 'straight' DNA fibres. In fact, it must be bent or curved in many of its functional roles, and is deformed from linearity in many of the dodecamer crystal structures, being typically bent by about 19°. There has been much speculation on the nature of DNA bending in such situations as chromatin (which has 145 base pairs wound around a histone protein core) and closed-circular DNA, prior to the determination of the chromatin crystal structure. Some sequences have an inherent tendency to bend independently of there being any protein present. The best-studied sequences are those comprising runs of adenines on one strand and thymines on the other (variously termed A tracts or A/T tracts).

Intrinsic DNA bending and curvature was discovered[88,89] in restriction fragments of kinetoplast DNA, which have extensive repeats of A tracts within them, each having about 5–7 base pairs. These fragments showed highly anomalous gel electrophoresis behaviour, with very slow migration through the gel, as if the DNA had a much higher molecular weight than was the case. Bending occurs when sequences or motifs such as 5′-AAAAAA are repeated in phase with the DNA helical repeat itself[90,91] that is every 10 or 11 base pairs. The phasing is critical since it forces the tract to be always on the same side of the helix, thereby enabling the individual small bends associated with each tract to add up to a significant amount of total bending. It has been estimated[92] that an individual $(A)_6$ tract bends a helix by about 17°–21°. A typical bending sequence, from kinetoplast DNA, is:

5′-···CCAAAAATGTCAAAAAATAGGCAAAAAATGCCAAAAAT···

Two distinct structural models have been suggested to account for these observations of bending. In one, a combination of gel retardation and NMR studies has been used,[93] resulting in a structural model that has bending towards the minor-groove direction of the helix. The compression and closure of the minor groove is in part a result of negative A•T base pair tilt within the A tract itself. In this, the 'junction' model, bending is caused by the abrupt change in structure when the A tract meets the more normal (straight) B-DNA intervening sequence, resulting in the helix axes of the two segments being at an angle to each other (Fig. 3.17a). Abrupt changes in either tilt or roll

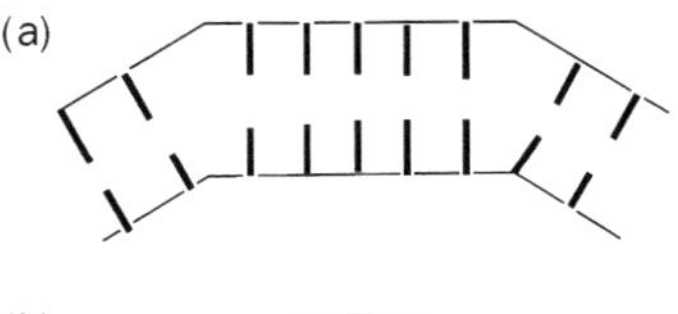

Fig. 3.17 Schematic representations of bending models for A tract DNA (a) junction model; (b) wedge model.

can produce this situation. Thus, in the junction model, the A tracts are themselves not bent. By contrast, the 'wedge' model[94] has continuous bending along the A tract (Fig. 3.17b) as a result of changes in roll at each dinucleotide step—this model has also been suggested for the continuous (induced) bending of general-sequence DNA when wrapped around histones in nucleosomes. The structural basis for bending has been controversial, there being continuing debate on the roles played by base pair propeller twist, roll, buckle and slide both within and flanking A tracts themselves. The advent of relevant crystal and NMR structures is providing structural data for distinguishing between these models, with increasing reliability for the various base and base pair morphological parameters, although no bent structure is as yet available that even approaches atomic resolution.

3.6.3 **Structures showing bending**

Several oligonucleotide crystal structures contain short dA•dT sequences that might be thought of as models for A tracts, as discussed in Sections 3.3.1 and 3.3.2. These have narrow minor grooves with ordered water molecules, high propeller twists for the A•T base pairs, together with, in two cases[33,34] bifurcated three-centre hydrogen bonds involving adjacent A•T base pairs (Fig. 3.12). Such hydrogen bonding was however, not observed[35] in the somewhat higher resolution structure of d(CGCAAATTTGCG). The $(A)_6•(T)_6$ tract in one crystal structure[33] is itself little bent, lending support to the junction model. All three crystal structures have an overall bending towards the major-groove direction, in contrast to the inference from gel studies, that A tracts are actually bent towards the minor groove. Crystal packing forces have been invoked to account for this discrepancy. More recently determined structures (see below) are probably better models for A tracts.

There is some evidence that A tracts do not inherently require three-centre A•T hydrogen bonds at all positions for curvature to be produced. For example, sequences having some inosine•cytosine (I•C) base pairs (these cannot form major-groove three-centre hydrogen bonds since inosine lacks a N6 group), still result in bent structures.[95] A combined crystallographic and gel electrophoresis study has been undertaken on a number of A tract model oligonucleotides with varying numbers of I•C base pairs in the tract.[96] This showed that high propeller twist and bending is maintained when there is isolated I•C substitution for A•T, but abolished for pure I tracts.

Interestingly, this analysis also suggested the existence of attractive three-centre NH $\cdots$ H interactions.[97]

The degree of bending in a DNA structure can be calculated with several of the helix morphology programmes outlined in Chapter 2. The use of normal vector plots is especially useful for visualizing bending.[98,99] This procedure uses the unit vectors perpendicular to the least–squares plane from a base pair, and enables both local and global bending to be analysed. The method was used to show that the Dickerson–Drew dodecamer[7,8,10] is bent by 19°, albeit in the major-groove direction. The crystal structure of the decamer d(CATGGCCATG) shows distinct curvature features. The helix is not straight but has a smooth 23° bend over the four central base pairs.[100] This intrinsic bend is consistent with a straight A tract model, although the roll compression of the major groove suggests a third type of model distinct from the wedge or junction models. The crystal structure[101] of the dodecamer d(ACCGAATTCGGT) is non-isomorphous with the Dickerson–Drew dodecamer, since it crystallizes in the triclinic space group P1 rather than the normal $P2_12_12_1$ one for dodecamers. This sequence is a biologically relevant one—it is the high-affinity A tract binding site of the E2 papillomavirus protein. There are three independent molecules in the crystallographic asymmetric unit, providing three independent views of the preferred structure of this sequence. All three are closely similar, with root-mean-square deviations ranging from 0.5–0.7 Å, strongly suggesting that their structure represents a global minimum. They have an average curvature of 9°, with very narrow minor grooves in the 5′-AATT regions and highly twisted A•T base pairs. The magnitude and direction of curvature is in accord with solution data on this sequence. The bending is produced by a combination of minor-groove compression in the A tract and major-groove compression of the flanking C•G base pairs. These are a consequence of the inherent features of the central octamer duplex sequence—in particular high propeller twists for the A•T base pairs and buckling at the A tract junctions.

The structure of the duplex formed by the non-self-complementary dodecamer d(GGCAAAAAACGG), bound to its complementary strand d(CCGTTTTTTGCC) has been analysed by NMR methods.[102] This A tract structure is bent by 19°, in accord with gel retardation measurements. The A•T base pairs in the A tract have high propeller twist, with most of the bending occurring in the short flanking C/G sequences. The A tract itself in this structure has a small amount (5°) of intrinsic bend. The junction bends have been ascribed to combinations of tilt and roll. NMR methods have also shown[103] that ammonium ions are bound in the A tract minor groove, and may play a role in the bending. These ions cannot be distinguished from water molecules in oligonucleotide structures by current X-ray methods.

It is striking that the macroscopic degree of bending observed in these structures is fully in accord with solution measurements and with the overall features of the junction model. Some features are common to all the structures such as a very narrow minor groove in the A tract coupled with high A•T base pair propeller twists. Yet the fine detail of how bending is achieved at and close to the junctions differ from one structure to another.[99] This suggests that there is inherent flexibility of the junctions to A tract structures, and that there are several equi-energetic bent arrangements that can be readily accessed, albeit in differing ways. This is supported by a survey of bending both in native oligomers and in DNA–protein complexes,[104] which also finds evidence for the straight A track model. There has been some debate as to whether, and to what extent, the bending in the crystallographically derived structures is affected by the near-universal use of 2-methyl-2,4-pentanediol (MPD) in crystallization experiments. Its role is to decrease the aqueous solubility of oligonucleotides. Electrophoresis mobility[105] and free-radical cleavage[106] experiments in the presence of MPD suggest that high concentrations do produce a change in A tract structure and decreased bending, although these 'low resolution' methods clearly cannot specify the molecular nature of the changes. On the other hand, the majority of the A tract model crystal structures *do* actually show solution-relevant bending, even though MPD may in some circumstances qualitatively affect its degree.[107]

3.6.4 The structure of poly dA•dT

The structure of poly dA•dT itself has been the subject of a number of studies. Fibre diffraction methods have defined a structure, the 'heteronomous' model[108] having distinct backbone conformations and sugar puckers for the two dA and dT strands. The results of more recent analyses,[109,110] although differing in detail, are consistent with this structure, with both strands having B-like conformations, albeit with a narrow minor groove. Propeller twists for the A•T base pairs are likely to be significant ($\sim 15°$), but not as high as in the earlier crystal structures, thus reducing the possibility of three-centre hydrogen bonding. Molecular dynamics studies[111–114] similarly indicate that this type of hydrogen bonding is not of primary importance in stabilizing oligo or poly dA•dT, but merely that it can occur when the geometric circumstances allow.

3.7 Concluding remarks

Fibre diffraction studies on DNA polynucleotides have long established the inherent flexibility of the double helix. These have also defined the major polymorphs in terms of A, B and Z duplexes. The fact that polynucleotide fibre structures can be readily interconvertible under appropriate environmental conditions has also shown us that they actually represent low-energy points along a continuum of structures. The very large number of oligonucleotide crystal structures now known are representative of this continuum, with the important caveat that it is sometimes not possible to distinguish between true fluctuations from the canonical mean, and experimental uncertainties. The aim of establishing a straightforward code for correspondence between sequence and structure remains elusive, and indeed may not exist. Instead, these structures are telling us much about DNA flexibility, its low-energy states and the pathways between them. The succeeding chapters show us how this flexibility can be exploited by interactions with other molecules, small and large.

References

1   Dickerson, R. E. *et al.* (1989). *The EMBO Journal*, **8**, 1.

2   Watson, J. D. and Crick, F. H. C. (1953). *Nature*, **171**, 737.

3   Arnott, S. (1999). In *Oxford Handbook of Nucleic Acid Structure* (ed. Neidle, S.), p. 1, Oxford University Press, Oxford.

4   Arnott, S., Chandrasekaran, R., Hall, I. H., Puigjaner, L. C., Walker, J. K., and Wang, M. (1982). *Cold Spring Harbor Symposium on Quantitative Biology*, **47**, 53.

5   Mahendrasingam, A., Forsyth, V. T., Hussain, R., Greenall, R. J., Pigram, W. J., and Fuller, W. (1986). *Science*, **233**, 133.

6   Arnott, S., Chandrasekaran, R., Birdsall, D. L., Leslie, A. G. W., and Ratliffe, R. L. (1980). *Nature*, **283**, 743.

7   Wing, R. M., Drew, H. R., Takano, T., Broka, C., Takana, S., Itakura, K., and Dickerson, R. E. (1980). *Nature*, **287**, 755.

8   Drew, H. R. and Dickerson, R. E. (1981). *Journal of Molecular Biology*, **151**, 535.

9   Kubinec, M. G. and Wemmer, D. E. (1992). *Journal of the American Chemical Society*, **114**, 8739.

10   Dickerson, R. E. and Drew, H. R. (1981). *Journal of Molecular Biology*, **149**, 761.

11   Dickerson, R. E., Goodsell, D. S., and Neidle, S. (1994). *Proceedings of the National Academy of Sciences USA*, **91**, 3579.

12   Liu, J. and Subirana, J. A. (1999). *Journal of Biological Chemistry*, **274**, 24749.

13   Minasov, G., Tereshko, V., and Egli, M. (1999). *Journal of Molecular Biology*, **291**, 83.

14   Johansson, E., Parkinson, G., and Neidle, S. (2000). *Journal of Molecular Biology*, **300**, 551.

15   Shui, X., McFail-Isom, L., Hu, G. G., and Williams, L. D. (1998). *Biochemistry*, **37**, 8341.

16   Tereshko, V., Minasov, G., and Egli, M. (1999). *Journal of the American Chemical Society*, **121**, 470.

17   Hamelberg, D., McFail-Isom, L., Williams, L. D., and Wilson, W. D. (2000). *Journal of the American Chemical Society*, **122**, 10513; Hamelberg, D., Williams, L. D., and Wilson, W. D. (2001). *Journal of the American Chemical Society*, **123**, 7745.

18   Trantírek, L., Urbášek, M., Štefl, R., Feigon, J., and Sklenár, V. (2000). *Journal of the American Chemical Society*, **122**, 10454.

19   Lomonossoff, G. P., Butler, P. J. G., and Klug, A. (1981). *Journal of Molecular Biology*, **149**, 745.

20   Drew, H. R. and Travers, A. A. (1984). *Cell*, **37**, 491.

21   Calladine, C. R. (1982). *Journal of Molecular Biology*, **161**, 343.

22   Calladine, C. R. and Drew, H. R. (1984). *Journal of Molecular Biology*, **178**, 773.

23   Suzuki, M., Amano, N., Kakinuma, J., and Tateno, M. (1997). *Journal of Molecular Biology*, **274**, 421.

24   Subirana, J. A. and Faria, T. (1997). *Biophysical Journal*, **73**, 333.

25   Gorin, A. A., Zhurkin, V. B., and Olson, W. K. (1995). *Journal of Molecular Biology*, **247**, 34.

26   Hunter, C. A. (1993). *Journal of Molecular Biology*, **230**, 1025.

27   Hunter, C. A. and Lu, X.-J. (1997). *Journal of Molecular Biology*, **265**, 603.

28   Packer, M. J. and Hunter, C. A. (2001). *Journal of the American Chemical Society*, **123**, 7399.

29   Schneider, B., Neidle, S., and Berman, H. M. (1997). *Biopolymers*, **42**, 113.

30   Packer, M. J. and Hunter, C. A. (1998). *Journal of Molecular Biology*, **280**, 407.

31   Packer, M. J., Dauncey, M. P., and Hunter, C. A. (2000). *Journal of Molecular Biology*, **295**, 85.

32   Yoon, C., Privé, G. G., Goodsell, D. S., and Dickerson, R. E. (1988). *Proceedings of the National Academy of Sciences USA*, **85**, 6332.

33   Nelson, H. C. M., Finch, J. T., Luisi, B. F., and Klug, A. (1987). *Nature*, **330**, 221.

34   DiGabriele, A. D. and Steitz, T. A. (1993). *Journal of Molecular Biology*, **231**, 1024.

35   Edwards, K. J., Brown, D. G., Spink, N., Skelly, J. V., and Neidle, S. (1992). *Journal of Molecular Biology*, **226**, 1161.

36   Webster, G. D., Sanderson, M. R., Skelly, J. V., Neidle, S., Swann, P. F., Li, B. F., and Tickle, I. J. (1990). *Proceedings of the National Academy of Sciences USA*, **87**, 6693.

37   Timsit, Y., Westhof, E., Fuchs, R. P. P., and Moras, D. (1989). *Nature*, **341**, 459.

38   Chiu, T. K. and Dickerson, R. E. (2000). *Journal of Molecular Biology*, **301**, 915.

39   Kielkopf, C. L., Ding, S., Kuhn, P., and Rees, D. C. (2000). *Journal of Molecular Biology*, **296**, 787.

40   Timsit, Y., Vilbois, E., and Moras, D. (1991). *Nature*, **354**, 167.

41   Timsit, Y. and Moras, D. (1991). *Journal of Molecular Biology*, **221**, 919.

42   Quintana, J., Grzeskowiak, K., Yanagi, K., and Dickerson, R. E. (1992). *Journal of Molecular Biology*, **225**, 379.

43   Lipanov, A., Kopka, M. L., Kaczor-Grzeskowiak, Quintana, J., and Dickerson, R. E. (1993). *Biochemistry*, **32**, 1373.

44   Goodsell, D. S., Grzeskowiak, K., and Dickerson, R. E. (1995). *Biochemistry*, **34**, 1022.

45   Heinemann, U., Alings, C., and Bansal, M. (1992). *The EMBO Journal*, **11**, 1931.

46   Yuan, H., Quintana, J., and Dickerson, R. E. (1992). *Biochemistry*, **31**, 8009.

47   Hartmann, B., Piazzola, D., and Lavery, R. (1993). *Nucleic Acids Research*, **21**, 561.

48   Schuerman, G. S. and van Meervelt, L. (2000). *Journal of the American Chemical Society*, **122**, 232.

49   Wang, A. H.-J., Fujii, A., van Boom, J. H., van der Marel, G. A., van Boeckel, S. A. A., and Rich, A. (1982). *Nature*, **299**, 601.

50   Shakked, Z., Guerstein-Guzikevich, G., Eisenstein, M., Frolow, F., and Rabinovich, D. (1989). *Nature*, **342**, 456.

51   Jain, S., Zon, G. and Sundaralingam, M. (1991). *Biochemistry*, **30**, 3567.

52   McCall, M., Brown, T., and Kennard, O. (1985). *Journal of Molecular Biology*, **183**, 385.

53   Eisenstein, M., Frolow, F., Shakked, Z., and Rabinovich, D. (1990). *Nucleic Acids Research*, **18**, 3185.

54   Lauble, H., Frank, R., Blücker, H., and Heinemann, U. (1988). *Nucleic Acids Research*, **16**, 7799.

55   Clark, G. R., Brown, D. G., Sanderson, M. R., Chwalinski, T., Neidle, S., Veal, J. M., *et al.* (1990). *Nucleic Acids Research*, **18**, 5521.

56   McCall, M., Brown, T., Hunter, W. N., and Kennard, O. (1986). *Nature*, **322**, 661.

57   Fairall, L., Martin, S., and Rhodes, D. (1989). *The EMBO Journal*, **8**, 1809.

58   Aboul-ela, F., Varani, G., Walker, G. T., and Tinoco, I, Jr. (1988). *Nucleic Acids Research*, **16**, 3559.

59   Verdaguer, N., Aymami, J., Fernández-Forner, D., Fita, I., Coll, M., Huynh-Dinh, T., *et al.* (1991). *Journal of Molecular Biology*, **221**, 623.

60   Frederick, C. A., Quigley, G. J., Teng, M.-K., Coll, M., van der Marel, G., van Boom, J. H., *et al.* (1989). *European Journal of Biochemistry*, **181**, 295.

61   Bingman, C. A., Zon, G., and Sundaralingam, M. (1992). *Journal of Molecular Biology*, **227**, 738.

62   Gao, Y.-G., Robinson, H. and Wang, A. H.-J. (1999). *European Journal of Biochemistry*, **261**, 413.

63   Finley, J. B. and Luo, M. (1998). *Nucleic Acids Research*, **26**, 5719.

64   Tippen, D. B. and Sundaralingam, M. (1997). *Biochemistry*, **36**, 536.

65   Eisenstein, M and Shakked, Z. (1995). *Journal of Molecular Biology*, **248**, 662.

66   Egli, M., Tereshko, V., Teplova, M., Minasov, G., Joachimiak, A., Sanishvili, R., *et al.* (1998). *Biopolymers*, **48**, 234.

67   Wahl, M. C. and Sundaralingam, M. (1997). *Biopolymers*, **44**, 45.

68   Malinina, L., Fernandez, L. G., Huynh-Dinh, T., and Subirana, J. A. (1999). *Journal of Molecular Biology*, **285**, 1679.

69   Ng, H.-L., Kopka, M. L., and Dickerson, R. E. (2000). *Proceedings of the National Academy of Sciences USA*, **97**, 2035.

70   Vargason, J. M., Henderson, K., and Ho, P. S. (2001). *Proceedings of the National Academy of Sciences USA*, **98**, 7265.

71   Ng, H.-L. and Dickerson, R. E. (2001). *Proceedings of the National Academy of Sciences USA*, **98**, 6986.

72   Wang, A. H.-J., Quigley, G. J., Kolpak, F. J., Crawford, J. L., van Boom, J. H., van der Marel, G., and Rich, A. (1979). *Nature*, **282**, 680.

73   Pohl, F. M. and Jovin, T. M. (1972). *Journal of Molecular Biology*, **67**, 375.

74   Rich, A. Nordheim, A., and Wang, A. H.-J. (1984). *Annual Reviews in Biochemistry*, **53**, 791.

75   Wang, A. H.-J., Quigley, G. J., Kolpak, F. J., van der Marel, G., van Boom, J. H., and Rich, A. (1981). *Science*, **211**, 171.

76   Egli, M., Williams, L. D., Gao, Q., and Rich, A. (1991). *Biochemistry*, **30**, 11388.

77   Fujii, S., Wang, A. H.-J., Quigley, G. J., Westerink, H., van der Marel, G., van Boom, J. H., and Rich, A. (1985). *Biopolymers*, **24**, 243.

78   Ban, C., Ramakrishnan, B., and Sundaralingam, M. (1996). *Biophysical Journal*, **7**, 1215.

79   Wang, A. H.-J., Hakoshima, T., van der Marel, G., van Boom, J. H., and Rich, A. (1984). *Cell*, **37**, 321.

80   Wang, A. H.-J., Gessner, R. V., van der Marel, G. A., van Boom, J. H., and Rich, A. (1985). *Proceedings of the National Academy of Sciences USA*, **82**, 3611.

81   Schneider, B., Ginell, S. L., Jones, R., Gaffney, B., and Berman, H. M. (1992) *Biochemistry*, **31**, 9622.

82   Basham, B., Eichman, B. F., and Ho, P. S. (1999). In *Oxford Handbook of Nucleic Acid Structure* (ed. Neidle, S.), p. 199, Oxford University Press, Oxford.

83   Rahmouni, A. R. and Wells, R. D. (1992). *Journal of Molecular Biology*, **223**, 131.

84   Schwartz, T., Rould, M. A., Lowenhaupt, K., Herbert, A., and Rich, A. (1999). *Science*, **284**, 1841.

85   Schwartz, T., Behike, J., Lowenhaupt, K., Heinemann, U., and Rich, A. (2001). *Nature Structural Biology*, **8**, 761.

86   Tullius, T. D. and Dombroski, B. A. (1985). *Science*, **230**, 679.

87   Rhodes, D. and Klug, A. (1981). *Nature*, **292**, 378.

88   Marini, J. C., Levene, S. D., Crothers, D. M., and Englund, P. T. (1982). *Proceedings of the National Academy of Sciences USA*, **79**, 7664.

89   Crothers, D. M. and Shakked, Z. (1999). In *Oxford Handbook of Nucleic Acid Structure* (ed. Neidle, S.), p. 455, Oxford University Press, Oxford.

90   Crothers, D. M., Haran, T. E., and Nadeau, J. G. (1990). *Journal of Biological Chemistry*, **265**, 7093.

91   Hagerman, P. J. (1990). *Annual Reviews in Biochemistry*, **59**, 755.

92   Koo, H.-S., Drak, J., Rich, J. A., and Crothers, D. M. (1990). *Biochemistry*, **29**, 4227.

93   Nadeau, J. G. and Crothers, D. M. (1989). *Proceedings of the National Academy of Sciences USA*, **86**, 2622.

94   Ulanovsky, L. and Trifonov, E. N. (1987). *Nature*, **326**, 720.

95   Diekmann, S., Mazzaralli, J. M., McLaughlin, L. W., von Kitzing, E., and Travers, A. A. (1992). *Journal of Molecular Biology*, **225**, 729.

96   Shatzky-Schwatrz, M., Arbuckle, N. D., Eistenstein, M., Rabinovich, D., Bareket-Samish, A., Haran, T. E., *et al.* (1997). *Journal of Molecular Biology*, **267**, 595.

97   Luisi, B., Orozco, M., Sponer, J., Luque, F. J., and Shakked, Z. (1998). *Journal of Molecular Biology*, **279**, 1123.

98   Goodsell, D. S. and Dickerson, R. E. (1994). *Nucleic Acids Research*, **22**, 5497.

99   Dickerson, R. E. (1998). *Nucleic Acids Research*, **26**, 1906.

100   Goodsell, D. S., Kopka, M. L., Cascioo, D., and Dickerson, R. E. (1993). *Proceedings of the National Academy of Sciences USA*, **90**, 2930.

101   Hizver, J., Rozenberg, H., Frolow, F, Rabinovich, D., and Shakked, Z. (2001). *Proceedings of the National Academy of Sciences USA*, **98**, 8490.

102   MacDonald, D., Herbert, K., Zhang, X., Polgruto, T., and Lu, P. (2001). *Journal of Molecular Biology*, **301**, 1081.

103   Hud, N. V., Sklenár, H., and Feigon, J. (1999). *Journal of Molecular Biology*, **286**, 651.

104   Young, M. A., Ravishankar, G., Beveridge, D. L., and Berman, H. M. (1995). *Biophysical Journal*, **68**, 2454.

105   Sprous, D., Zacharias, W., Wood, Z. A., and Harvey, S. C. (1995). *Nucleic Acids Research*, **23**, 1816.

106   Ganunis, R., Guo, H., and Tullius, T. D. (1996). *Biochemistry*, **35**, 13729.

107   Dickerson, R. E., Goodsell, D., and Kopka, M. L. (1996). *Journal of Molecular Biology*, **256**, 108.

108   Arnott, S., Chandrasakaran, R., Hall, I. H., and Puigjaner, L. C. (1983). *Nucleic Acids Research*, **11**, 4141.

109   Chandrasakaran, R. and Radha, A. (1992). *Journal of Biomolecular Structure and Dynamics*, **10**, 153.

110   Aymami, J., Coll, M., Frederick, C. A., Wang, A. H.-J., and Rich, A. (1989). *Nucleic Acids Research*, **17**, 3229.

111   Fritsch, V. and Westhof, E. (1991). *Journal of the American Chemical Society*, **113**, 8271.

112   Sherer, E. C., Harris, S. A., Soliva, R., Orozco, M., and Laughton, C. A. (1999). *Journal of the American Chemical Society*, **121**, 5981.

113   Sprous, D., Young, M. A., and Beveridge, D. L. (1999). *Journal of Molecular Biology*, **285**, 1623.

114   Strahs, D. and Schlick, T. (2000). *Journal of Molecular Biology*, **301**, 643.

Further reading

Arnott, S. (1970). *Progress in Biophysics and Molecular Biology*, **6**, 265.

Berman, H. M. (1997). *Biopolymers*, **44**, 23.

Berman, H. M., Gelbin, A., and Westbrook, J. (1996). *Progress in Biophysics and Molecular Biology*, **66**, 255.

Dickerson, R. E. (1992). *Methods in Enzymology*, **211**, 67.

Dickerson, R. E. (1999). In *Oxford Handbook of Nucleic Acid Structure* (ed. Neidle, S.), Oxford University Press, Oxford.

Grzeskowiak, K. (1996). *Chemistry and Biology*, **3**, 785.

Shakked, Z. and Rabinovich, D. (1986). *Progress in Biophysics and Molecular Biology*, **47**, 159.

Wahl, M. C. and Sundaralingam, M. (1997). *Biopolymers*, **44**, 45.

4

DNA–DNA recognition: non-standard and higher order DNA structures

4.1 Mismatches in DNA

4.1.1 General aspects

Base pairing in DNA should not be considered solely in terms of Watson–Crick hydrogen bonding. Even though 16 distinct arrangements for A•T and G•C base pair hydrogen bonding are in principle possible, not all have actually been observed. The Hoogsteen and reverse Hoogsteen pairs (Fig. 4.1a,b) have been found in several crystal structures of adenine•thymine and adenine•uracil complexes. These arrangements involve atoms N6 and N7 of adenine rather than the N1 and N6 atoms of Watson–Crick hydrogen-bonding, so Hoogsteen hydrogen bonding with thymine is with the major-groove edge of the adenine base, at right angles from the one that forms Watson–Crick pairs in B-DNA. Thus Hoogsteen pairing implies that the adenine base has to be in a *syn* glycosidic angle conformation if the resulting base pair is to be present in an anti-parallel DNA duplex.

Mismatched base pairing in DNA can arise naturally during replication. The possible arrangements are: G•A, G•T, A•A, G•G, T•T, C•C, T•C or A•C. If left unchecked, these mismatches will lead to mutations and nonsense or incorrect gene products following replication. The cell, however, has evolved a number of enzymatic repair mechanisms to first, recognize mispairs whose efficiency depends on the nature of the mispairing, and often, on sequence context. The details of these mechanisms have been extensively

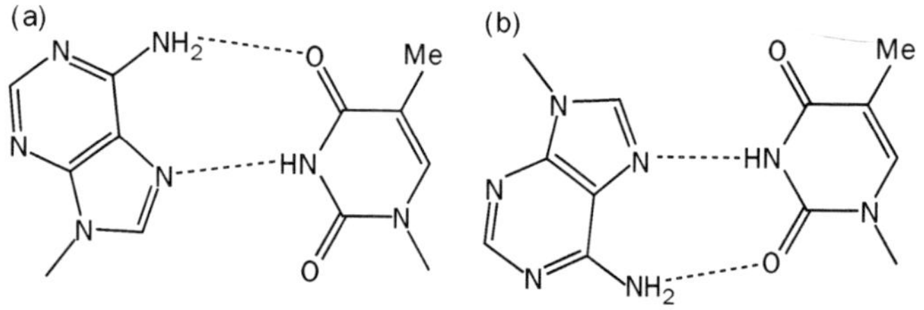

Fig. 4.1 Structures of Hoogsteen and reverse Hoogsteen A•T mismatched base pairs.

studied in *E. coli*, but are less well understood in higher organisms, at least in structural terms. In general, recognition of a mismatch is followed by duplex unwinding and excision of the mispair. A considerable number of structural studies have been reported on proteins involved in DNA repair, together with their complexes with various damaged DNAs.

Structural studies on DNA mispairs themselves have addressed the questions of:

1. the nature of the hydrogen-bonding involved in the pairings
2. how these affect the local and global structure of duplex DNA
3. the possible relationships between structure, sequence context of the mispair, and efficiency of mismatch repair.

Mismatches generally destabilize duplex DNA. (This can be shown by the extent to which the temperature at which the double helix⇔coil transition occurs, is decreased). Of the eight mispairs listed above, the A•G one has been the best-studied and is the one that is focused upon here. Structural studies have also been reported on oligonucleotides in A, B and Z forms with G•T, I(inosine)•T, U•G, I•C and I•A base pairs.

Mutagenesis (which can eventually lead to carcinogenesis) can occur when external agents such as methylating compounds produce covalent adducts with DNA bases. Until recently, little structural work has been reported on covalently modified oligonucleotides, largely because of the difficulties involved in producing sufficient pure quantities of adducts for X-ray and NMR studies. These difficulties have now been largely overcome, and an increasing number of structural studies (especially using NMR techniques), have now been performed on oligonucleotides incorporating, for example, methylated bases[1] and carcinogens such as benzo[a]pyrene. Of particular interest has been the mutagenic methylation of the O6 atom in guanine, which is produced by a variety of environmental and chemotherapeutic agents.

4.1.2 **Purine : purine mismatches**

Crystal structures have been determined for a number of self-complementary oligonucleotide sequences with A•G base pairs (Table 4.1). These show a variety of base pairings

Table 4.1 Some purine : purine mismatched oligonucleotide crystal structures, with the mismatched base pairs shown in bold. Distances are in Å

Sequence	Glycosidic angles	C1′ ⋯ C1′ distances	Reference
5′-CGC**GAATTA**GCG GCGA**TTAA**GCGC	G(*anti*)•A(*syn*)	10.6, 10.8	2
5′-CGC**AAATT**GGCG GCGG**TTAA**ACGC	A(*anti*)•G(*syn*)	10.8 (average)	3
5′-CGC**AAGCT**GGCG GCGG**TCGA**ACGC	A(*syn*)•G(*anti*)	10.7 (average)	4
5′-CCAA**GA**TTGG GGTT**AG**AACC	A(*anti*)•G(*anti*)	12.5	5
5′-CGC**GAATTG**GCG GCGG**TTAA**GCGC	G(*anti*)•G(*syn*)	10.7, 11.2	6
5′-CC**GAATGA**GG GG**AGTA**AGCC	A(*anti*)•G(*anti*)	8.7	12

(Fig. 4.2). The arrangement with both nucleosides in an *anti* conformation necessarily forces the inter-strand C1′ · · · C1′ distance to be further apart than in a standard B-DNA Watson–Crick base pair (10.4 Å). This *anti, anti* form (Fig. 4.2a) has only been observed in one crystal structure,[3] where there are two consecutive A•G base pairs. Presumably the bulge in the helix produced by their excessive C1′ · · · C1′ distances is relieved by the two A•G base pairs being well stacked on each other. For isolated purine : purine base pairs, it is easier to achieve normal C1′ · · · C1′ distances, and hence greater helix stability, by having one base *anti* and the other in a *syn* arrangement (Fig. 4.2b). The base pairing is of the Hoogsteen type. Energetic considerations suggest that a *syn* conformation would be preferred for the guanine rather than the adenine base, but Table 4.1 shows that this is only sometimes the case. All of the A•G-containing oligonucleotide structures have B-form structures, with generally standard geometries and few distortions from ideality, except that the mispairs tend to be poorly stacked with their neighbouring base pairs. Detailed examination of the stacking patterns has shown that the particular glycosidic conformation adopted by a mismatched base is such as to maximize the stacking, and so overrides the slight energy penalty of adenine being in a *syn* conformation. The nature of the sequence surrounding the mismatch is important, with most of the established structures following the rule that when the mispaired base on the first strand is at the centre of the three-residue sequence 5′-Py–**Pu**–Pu, then the central purine is in an *anti* conformation. An oligonucleotide with guanine : guanine base pairs has also been found[6] to have a B-like structure and an *anti, syn* arrangement for the mismatched bases (Fig. 4.3). It appears that this unusual mismatch is associated with B_{II} backbone phosphate conformations (see Section 3.3.1), which have also been documented for some adjacent G•A mismatches.[7] The B_{II} conformation results in altered phosphate groups intrastrand distances; these changes from standard helical geometry could act as recognition signals for repair nucleases, to identify and excise the mispairs at these points.

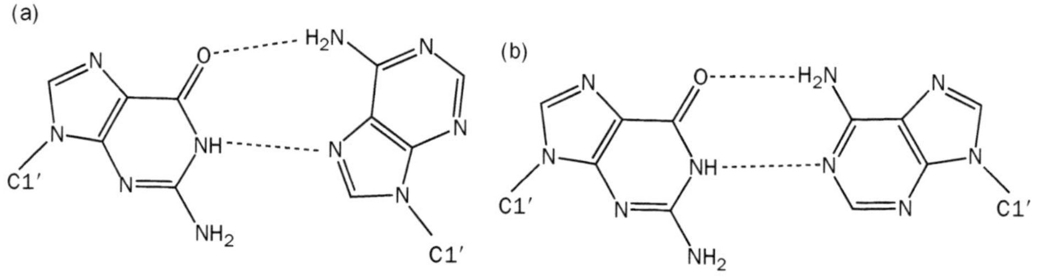

Fig. 4.2 Structures of G(*anti*)•A(*anti*) and G(*anti*)•A(*syn*) mismatches.

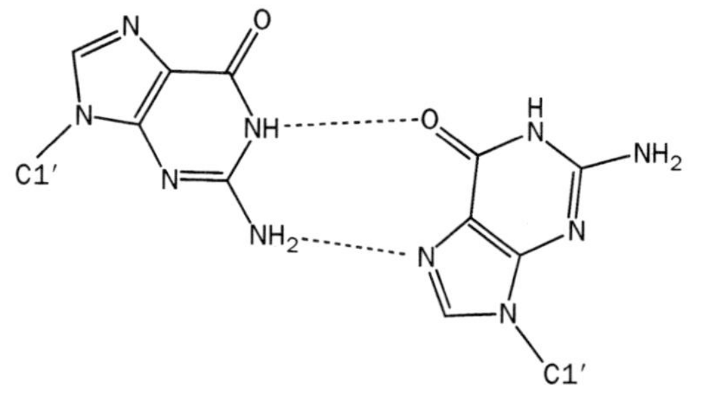

Fig. 4.3 A G(*anti*)•G(*syn*) base mismatch, as observed in a mismatch dodecamer crystal structure.[6]

NMR studies of G•A-containing oligonucleotides have shown that the pattern of glycosidic angles seen in the crystal structures is not always maintained in solution, and some arrangements can be readily converted into others with changes in pH.[8] Thus, the duplex of d(CGCAAATTG-GCG)$_2$ adopts an A(*anti*)•G(*anti*) arrangement at high pH,[9] yet an A$^+$(*anti*)•G(*syn*) one at low pH. It should be borne in mind that the typical NMR experiment uses somewhat different oligonucleotide solution conditions compared to those in crystallization trials, with higher salt concentrations in the former and high hydrophobic-type solvent concentrations in the latter. These differences are likely to have an effect on the conformational equilibria of equi-energetic mispairs.

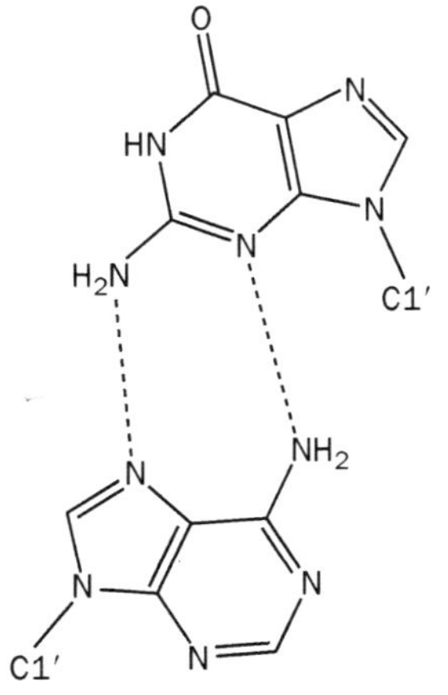

The variability of G•A base pairing arrangements (and hence their stability) is largely dependent on the optimization of base-stacking interactions, and hence on the nature of adjacent bases in a sequence. This is graphically illustrated in the sequence d(ATGAGCGAATA)$_2$, with four G•A base pairs.[10] NMR and molecular modelling show that these have N7 · · · N2 and N3 · · · N2 hydrogen bonds, with only this arrangement being capable of stabilization by stacking interactions with adjacent bases. Thus, in general, the sequence 5'-Py—GA—Pu forms a particularly stable mispaired unit,[11] regardless of the details of the hydrogen bonding involved. This stability is seen to be retained in the high-resolution crystal structure[12] of the duplex formed by the sequence d(CCGAATGAGG), which is a model for centromeric DNA, with the repeated sequence d(GAATG)$_n$. These tandem G(*anti*)•A(*anti*) mismatches (Fig. 4.4) in centromeric DNA have been termed 'sheared' on account of their side-by-side nature. The notable feature of this crystal structure is the extensive interstrand G/G and A/A base stacking, which helps to stabilize the unusual G•A base pairing. These features have also been observed in NMR structures of analogous sequences.[13,14]

Fig. 4.4 The tandem G(*anti*)•A(*anti*) mismatch.

4.1.3 **Alkylation mismatches**

Methylation at the O6 position of a guanine base in duplex DNA induces the resulting destabilizing G(OMe)•C base pair to undergo a transition mutation to a G(OMe)•T base pair when the DNA is replicated. Methylation of the O6 carbonyl group changes the C6—O6 bond from a double bond to a single one, and hence alters the overall pattern of tautomerism in the guanine ring system, with atom N1 no longer having an attached hydrogen atom. NMR studies on the G(OMe)•C mispair in an oligomer duplex[15] have indicated that both bases have *anti* glycosidic angles, in an overall B type helix, with 'wobble' hydrogen bonding solely between N2(G) and O2(C), and with the O6 methyl group oriented towards the cytosine (Fig. 4.5a). The crystal structure of a (Z-DNA) oligonucleotide duplex containing this mispair[16] shows an arrangement that is more compatible with standard Watson–Crick hydrogen-bonding—distances O6(G) · · · N4(C) and N1(G) · · · N3(C) are normal hydrogen-bonding ones (Fig. 4.5b). This suggests that even though the crystals were grown at pH 7.0, atoms N1(G) or N3(C) have acquired a proton so that partial Watson–Crick hydrogen bonding can occur, which is still less stable than a standard G•C base pair. The various possible arrangements for A•G mismatches have been described in the previous section. When the guanine in this mismatch is methylated, the number of possible hydrogen-bonded base pair arrangements becomes even greater,[17] although sequence context and/or a requirement for adenine protonation will exclude the majority from consideration in specific instances. The crystal structure[17] of a dodecanucleotide duplex with two such A•G(OMe) mispairs, shows an arrangement with two hydrogen bonds and A(*syn*)•G(*anti*) glycosidic angles (Fig. 4.5c). A common theme in this and many other mispair crystal structures is the location of water molecules hydrogen-bonded to (and often bridging between) the bases involved in the mismatches.

The G(OMe)•T base pair, which also destabilizes a duplex, has been observed to have Watson–Crick-type hydrogen bonding in two B-DNA dodecamer crystal structures,[18,19] with two hydrogen bonds (Fig. 4.5d). The overall shape of this base pair is close to that of a standard G•C one, whereas the G(OMe)•C one is rather less so, especially in the wobble arrangement (Fig. 4.5a). This difference may explain the observed preference for G(OMe)•T base pairs to be incorporated into DNA during replication and not to be

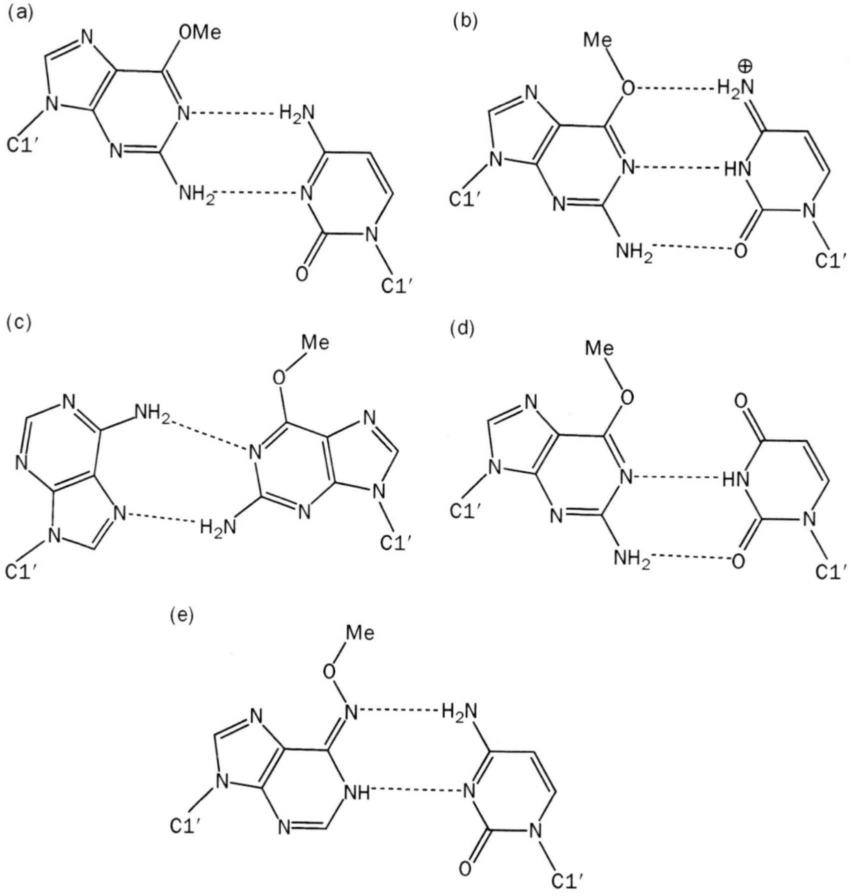

Fig. 4.5 (a) The G(OMe)•C wobble base pair; (b) the G(OMe)•C Watson–Crick base pair; (c) the A•G(OMe) base pair; (d) the G(OMe)•T base pair; (e) the A(OMe)•C Watson–Crick base pair.

Table 4.2 DNA mismatch crystal structures

Base pair mismatch	NDB number
G•T	BDL009
A•C	BDL011
G•A	BDL012
A•G	BDL014
A•G	BDL022
G•G	BDL046
A(OMe)•C	BD0009
A(OMe)•T	BD0010
G(OMe)•T	BDLB58
G•A(8-oxa)	BDLB33
G(OMe)•C	ZDFB21

readily recognized by repair enzymes. Mutagens such as hydroxylamine can methoxylate the N6 position of adenine. The crystal structure[20] of a dodecamer with two methoxyadenine residues shows that they form analogous Watson–Crick-like base pairs (Fig. 4.5e) with a cytidine on the opposite strand. In order for this to occur, the modified adenines must be in the imino form, as shown, enabling it to mimic a guanine, with the result that cytidine rather than thymine is mis-incorporated during replication. Thymine opposite this adenine lesion also results in Watson–Crick-like base pairing.[21]

Oxidative damage to bases can occur by free radical attack, especially at guanine sites, where the primary product is 7,8-dihydro-8-oxa-guanine. The crystal structure of a duplex containing this lesion[22] shows that the standard Watson–Crick G•C hydrogen bonding arrangements are unimpaired, and alternative G(*syn*) conformations are not involved, in accord with NMR studies. This positions the 8-oxa group in the major groove, where it is available for recognition by DNA repair enzymes (Table 4.2).

4.2 **DNA triple helices**

4.2.1 **Introduction**

The formation of a triple helix by two pyrimidine strands and one purine strand, was discovered[23] soon after the structure of the double helix itself was determined, when solutions of poly(A) and poly(U) were mixed in appropriate proportions, forming a 1 : 1 : 1 three-stranded polynucleotide complex, poly(U•AU). A molecular model for this novel helix was proposed on the basis of fibre diffraction data[24] for both deoxy and ribo-polynucleotides. This indicated that the triple helix is right-handed. The fibre diffraction data is consistent with an (adenine) purine strand Watson–Crick hydrogen-bonding to a (thymine or uracil) pyrimidine one, and a third (thymine or uracil) pyrimidine strand Hoogsteen hydrogen-bonding to the purine strand and parallel to this strand (Figs 4.6a and 4.7). This hydrogen-bonding arrangement is termed a base triplet and the triple helix itself is often termed a triplex. We adopt the convention here for the triplet X•YZ, that YZ represents the Watson–Crick hydrogen-bonded base pair, with base X being in the third strand and hydrogen-bonded in some way to base Y.

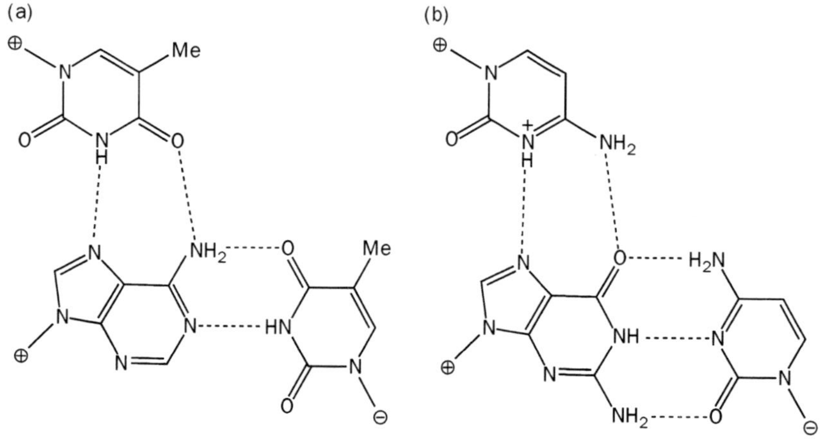

(a) (b)

Fig. 4.6 The T•AT and C⁺•GC triplex hydrogen-bonding arrangements in a Py–Pu–Py triple helix.

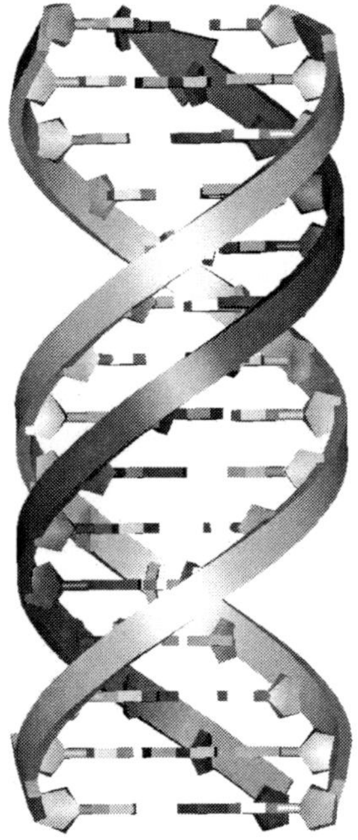

Fig. 4.7 Schematic view of a DNA parallel triple helix, with the third (pyrimidine) strand shown shaded.

A guanine and two cytosine-containing strands can also form a parallel triple helix, which, therefore, has the same polarity of strands as the T•AT triplex. However, now the third-strand cytosine is required to be protonated at the N3 position in order for two hydrogen bonds to be formed and so effective Hoogsteen hydrogen bonding can take place (Fig. 4.6b). The C⁺•GC triplex is isosteric with the T•AT one. Protonation of a cytosine base at N3 can only take place at about pH 5.0, well below physiological pH. This places a potentially severe practical limitation on C⁺•GC triplex formation in biological systems (see below). Substitution of a methyl group or bromine at the 5-position of cytosine results in some stabilization of triplex formation for C⁺•GC-containing sequences at *c.* pH 7, (i.e. around physiological pH). The precise reasons for this are not clear. It is possible that there is an increased hydrophobic contribution to third-strand binding when there are these substituents at the 5-position of cytosine, rather than there being a shift in the pK_a of the N3 atom of cytosine. Triple helices can be stable under some circumstances with both cytosines and thymines in the third strand, that is, with mixed pyrimidine sequences. These are further discussed below.

The triple helix phenomenon was for many years no more than a laboratory curiosity, until the findings that stretches of triple helix could be formed by oligonucleotides of appropriate sequence binding to long duplex DNA molecules (Fig. 4.8a). Such a sequence (5′-TTCTTTTCTTCTTTCTTTTT), with a covalently attached strand cleavage agent, has been used to bind to and cleave a unique site on a yeast chromosome,[25] demonstrating the exceptionally high specificity for a target duplex sequence that triplex formation can achieve. Despite this specificity, triple helices are inherently

(a)

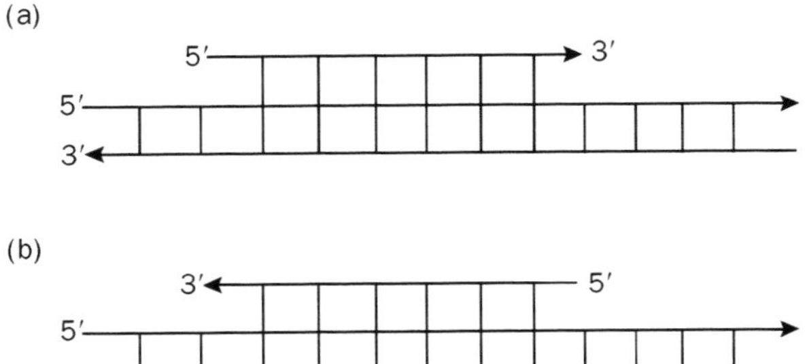

(b)

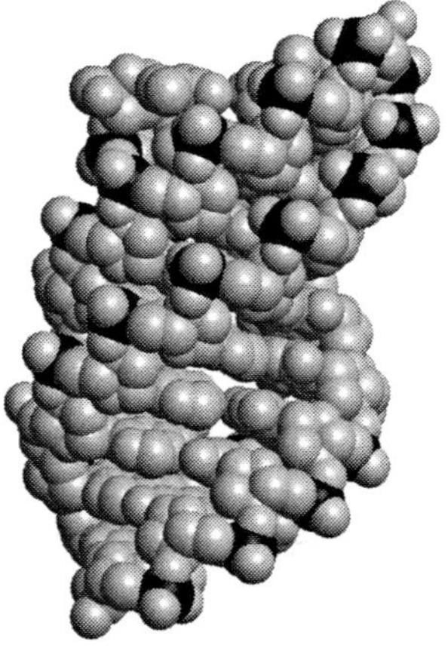

Fig. 4.8 Schematic view of the two categories of triple helix formed by an oligonucleotide binding to a long stretch of duplex DNA, with strand directions indicated by arrows. The third strand is shown (a) in a parallel orientation and (b) in an antiparallel orientation to the first duplex strand, which is conventionally a purine one.

readily disrupted as a consequence of the proximity of the third strand and its negatively charged phosphate groups to the two duplex strands.

4.2.2 Structural studies

The model derived from the original fibre diffraction studies[24] suggested that the third strand in this triple helix occupies the major groove of the purine–pyrimidine double helix. This key feature has been retained in all subsequent models and ultimately verified by both NMR and single-crystal studies. The parallel triple helix in the fibre diffraction model has some (though not all) features of classic A-DNA (Fig. 4.9). It has bases significantly (~3.5 Å) displaced from the helix axis and an average helical twist of ~30°. The duplex minor-groove width of 10.7 Å is almost identical to the A-DNA one of 11.0 Å. However its wide major-groove width, of 9.8 Å, which is necessary to accommodate the third strand, is much greater than in standard duplex A-DNA. All deoxyribose sugars in this model have C3′-*endo* pucker.

Fig. 4.9 The A-form parallel triple helix, as suggested from fibre diffraction studies,[24] shown as a space-filling structure. Phosphorus atoms are shaded black, and the third Hoogsteen strand is at the front of this view, nestling in the enlarged major groove.

   Triple helices can be formed by relatively short oligonucleotides in solution, such as $d(A)_{12} \cdot 2d(T)_{12}$, and a number of NMR studies have been performed[26–29] on such systems. Intramolecular triplexes formed by a single strand of appropriate sequence folding back on itself tend to be more stable and have been especially well studied. The sugar residues in parallel triplexes tend to have B-DNA type C2′-*endo* sugar puckers, whereas the helical twists are closer to those of A-DNA. Overall the parallel triple helices derived from NMR data have morphology closest to that of B-type duplex DNA, with the differences in, for example, x-displacement, being due to the need to expand the size of the major groove to accommodate the third strand (Fig. 4.7). The third DNA strand in a parallel triplex can be replaced by an RNA one, with some increase in stability. This hybrid triplex still retains its overall B-type character,[30] in spite of the presence of the RNA strand, with sugar puckers of the DNA component in the C2′-*endo* range and an x-displacement of -0.7 Å.

   A number of molecular mechanics and dynamics simulation studies have been performed on triple-helical DNAs. These studies have not produced a single type of model, but rather have conflicting features such as sugar puckers and helix morphologies.[31,32]

This is a reflection of several factors that are common to all nucleic acid simulations, but are probably most marked for triplexes because of the difficulty of being able to adequately treat the electrostatics of three negatively charged DNA strands. This has now been overcome by the use of the particle-mesh Ewald algorithm. Differences in sugar pucker are in large part a result of the earlier use of different force-fields with subtly different pucker preferences. The increasing ability to undertake long-time-scale, electrostatically stable simulations, together with improved electrostatic treatment, has recently resulted in simulated parallel triplex B-type structures that have many features in common with experimental data.[33]

Until recently there has been no single crystal structure, in spite of much effort, which tended to produce at best crystals with excellent external morphology but little internal order.[34] Several crystal structures of duplex DNA oligonucleotides with overhanging ends, have revealed the details of triplet arrangements of bases in the crystal lattice. For example, the structure of d(GCGAATTCG) involves Hoogsteen G•GC base triplets formed by interactions between adjacent molecules in the crystal.[35] However, these structures do not provide information about the morphology of triple helices themselves. A triple-stranded arrangement has been found in the structure[36] of a complex between an 18-mer oligopyrimidine foldback peptide nucleic acid (PNA) sequence, and a 9-mer oligopurine sequence (Fig. 4.10). (PNA molecules are nucleic acid mimics in which the sugar-phosphate linkage between successive bases has been replaced by an uncharged 2-aminoethylglycine unit with closely similar spacial properties to the natural linker (Fig. 4.10). They form very stable hybridization complexes both with other PNA molecules and with DNA oligonucleotides).[37,38] The PNA–DNA triplex crystal structure (Fig. 4.11 and Table 4.3) has features distinct from those of either A- or B-DNA, with 16 bases per turn and a large x-displacement, of 6.8 Å. All sugars of the DNA strand have C3′-*endo* puckers. All of the 10 T•AT and eight C+•GC base triplets in this structure have the expected Hoogsteen geometry. The overall morphology of this triplex, which has been

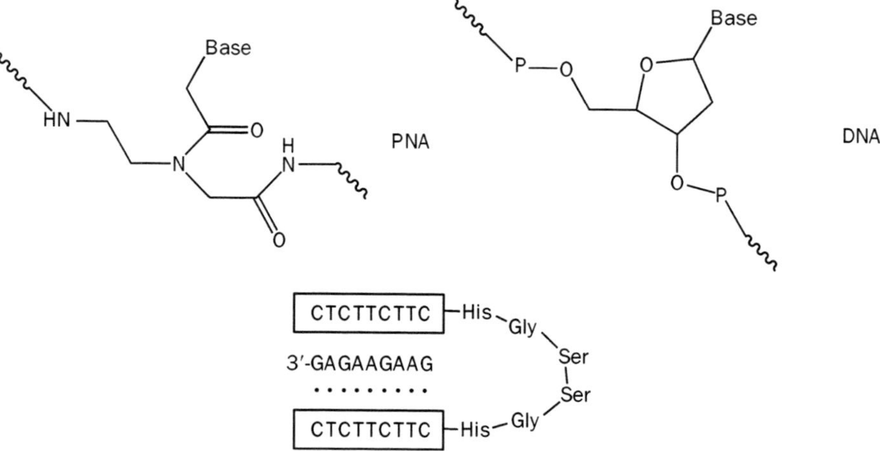

Fig. 4.10 The molecular structure of the PNA repeating unit, compared to a DNA backbone. Also shown are the sequences of DNA and PNA strands in the crystal structure of the foldback PNA–DNA triple helix.[36] The PNA strands are shown highlighted in boxes.

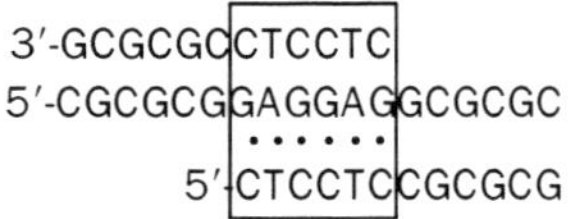

Fig. 4.12 The sequences co-crystallized to form a parallel DNA triple helix,[39] with the triple helical segment highlighted.

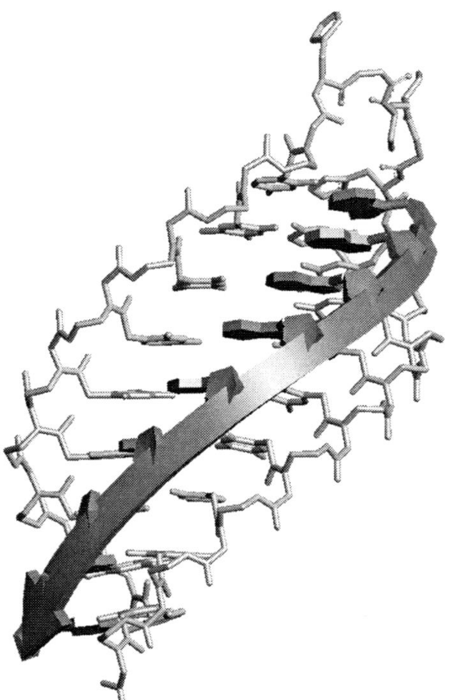

Fig. 4.11 The crystal structure of the PNA–DNA triple helix, with the DNA strand highlighted.[36]

Table 4.3 Crystal structures of nucleic acid triple helices

Triple helix type	NDB code no.
PNA•DNA–PNA	PNA001
DNA•DNA–DNA	BD0017

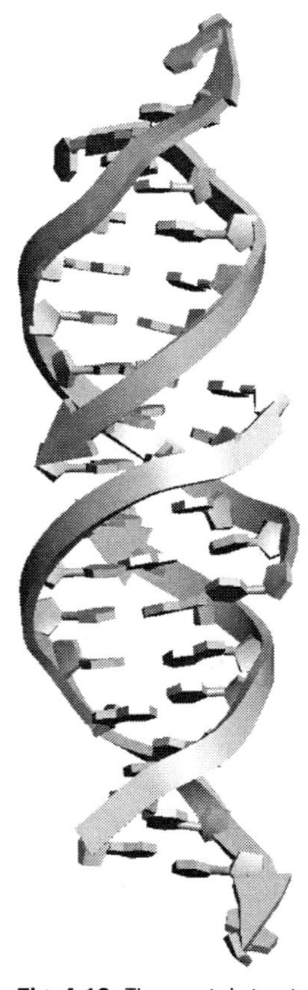

Fig. 4.13 The crystal structure of the parallel DNA triple helix.[39]

termed the P-form, is governed by the particular requirements of the PNA strands, which are known to prefer to adopt A-type geometries.

A true DNA triple helix has been crystallized,[39] using a construct of three overlapping oligonucleotide sequences, such that there is a central symmetric segment of six base triplets forming a short stable triplex region, flanked by a hexamer duplex at each end (Fig. 4.12). Crystals were grown at pH 5.0, in view of the dominance of C^{+}•GC triplets in the sequence. This crystal structure (Fig. 4.13), at 1.8 Å resolution, confirms all the general features of triplexes inferred from earlier modelling and NMR studies. In particular, the short distances (2.7 Å) between N3 of a cytosine and N7 of a guanine in a C^{+}•GC triplet confirm the existence of a second hydrogen bond in this C•G Hoogsteen base pair, and thus of cytosine N3 protonation (Fig. 4.14). Most of the sugars have C2'-*endo* pucker. This feature, combined with the low base inclination of 6° in the triplex region and the *x*-displacement

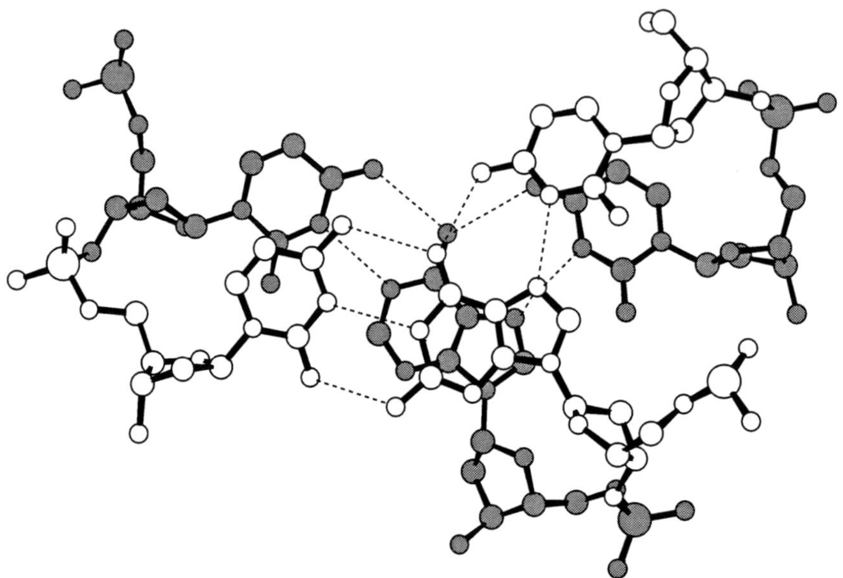

Fig. 4.14 View onto two successive base triplets in the triple-helix crystal structure.[39]

of -2 Å, suggests that the triplex part of the structure could be categorized closer to the B-type family than A-type. However, as with all triplex structures, other features do not accord with this classification, and the structure is best not classified in such terms.

4.2.3 Anti-parallel triplexes and non-standard base pairings

Triple helices can also be formed with the third strand having solely purines, so that the resulting helix has Pu•PuPy triplets of bases.[40] Such a triple helix forms most readily with G•GC base triplets, having a G•G hydrogen bonding arrangement that is likely to be as shown in Fig. 4.15a, with the third, purine strand being anti-parallel to the purine strand of the duplex. This arrangement is pH-independent, contrasting with the C$^+$•GC triple helix. The overall stability of anti-parallel triplexes is frequently greater than that of parallel ones. The guanosine nucleoside on the third strand of the G•GC base triplet can have either *anti* or *syn* arrangements (Fig. 4.15a,b). Both NMR[29,41] and molecular modelling studies[42] suggest that the *anti* one is actually preferred, though no crystal structures of anti-parallel triplexes have been reported to date.

There have been numerous attempts to extend the sequence restrictions on triple-strand recognition[43] using natural DNA bases. There are two over-riding problems:

1. the need to maximize base $\cdots$ base triplet hydrogen bonding;
2. the distortions in DNA structure that are introduced when inserting one or a few purines in an oligopyrimidine third strand (and vice versa). These distortions reduce intrastrand base stacking and are probably a major contributor to the instability of many possible base triplets.

However, some particular instances do work. For example, a thymine in a pyrimidine third strand can be replaced by a guanine, with the resultant G•TA base triplet (Fig. 4.16a)

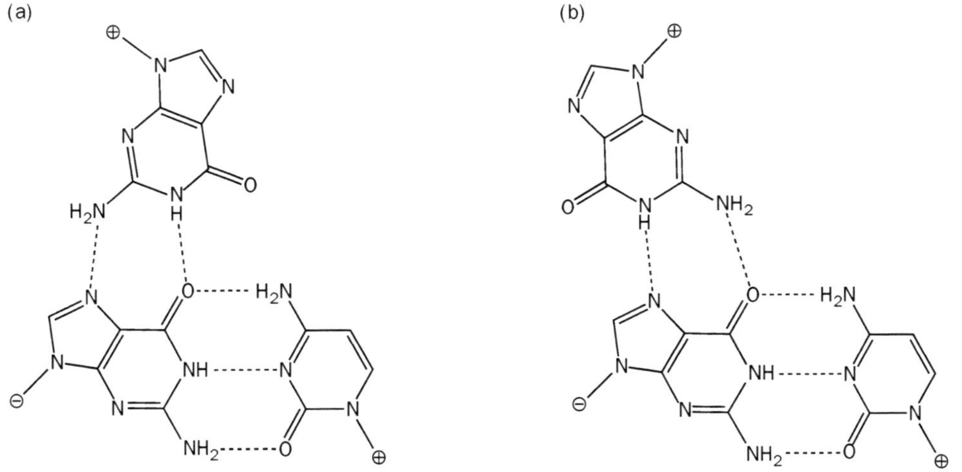

(a)

(b)

Fig. 4.15 (a) Anti-parallel and (b) parallel G•GC base triplet hydrogen bonding arrangements.

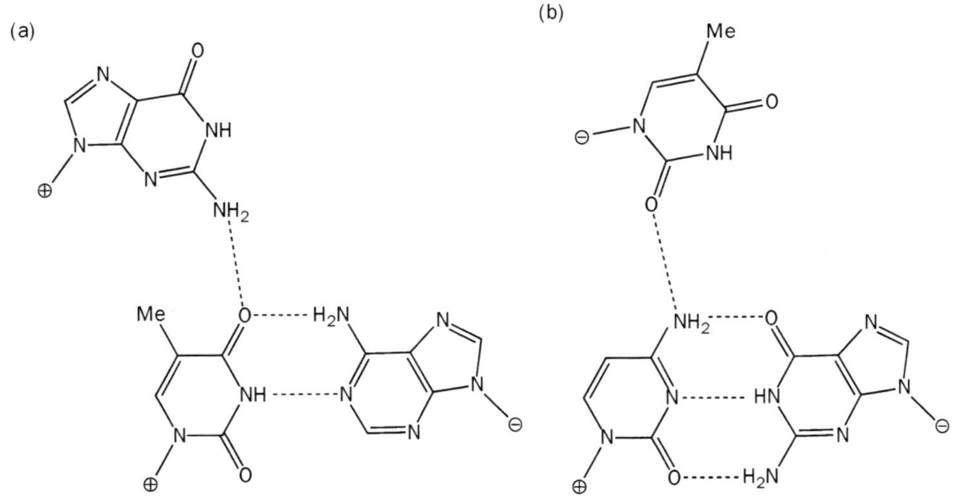

(a)

(b)

Fig. 4.16 (a) The G•TA base triplet; (b) the T•CG base triplet.

being relatively stable.[44] However the distortions that this is likely to produce in the DNA backbone suggest a limit on the number of such triplets that can be tolerated within a sequence. Almost all other possible triplet mismatches are highly destabilizing to triple helix formation[45,46] such that triplex formation for an otherwise stable sequence can be abolished by even a single such mismatch. The nature of the flanking sequences is generally important for stability of all types of non-standard triplet sequence. There are several mismatches, such as T•CG (Fig. 4.16b), that may be stable in some flanking sequence circumstances, but clearly do not work in others. Table 4.4 details the 'code' for triple-strand recognition.

Table 4.4 Relative stabilities of all 16 possible base triplets. Adopted from Soyfer and Potaman, 1996

Triplet	Stability
$^+$C•GC	++++
$^+$A•GC	++
G•GC	+
T•GC	++
C•AT	+
A•AT	+
G•AT	+
T•AT	++++
C•CG	+
A•CG	−−
G•CG	+
T•CG	++
C•TA	−−
A•TA	−−
G•TA	+++
T•TA	+

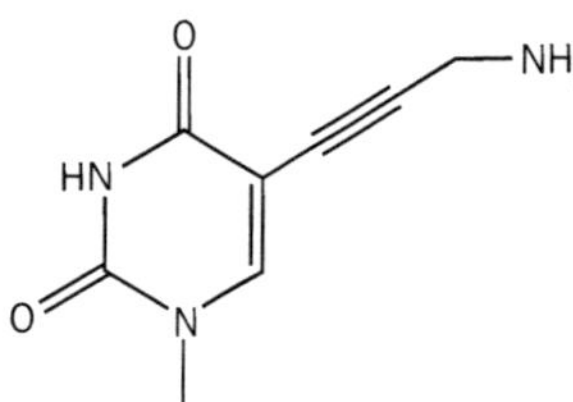

Fig. 4.17 The structure of 5-(1-propargylamino)-2′-deoxyuridine.[48]

More general mixed sequence recognition of all four Watson–Crick base pairs in a duplex (i.e. C•G, G•C, A•T and T•A) is almost certainly only achievable with non-standard 'designer' bases.[47] These may be capable of being incorporated into a general-sequence oligonucleotide and hybridizing with a target duplex with high affinity. The use of positively charged analogues of thymine, for example, 5-(1-propargylamino)-2′-deoxyuridine (Fig. 4.17), results in a 100-fold increase in binding affinity,[48] possibly as a result of favour able electrostatic interactions with adjacent phosphate groups.

The requirement for the C$^+$•GC triplet to be protonated is a major limitation when selecting potential genomic targets for triplex recognition (see below). This is especially constraining with contiguous C$^+$•GC triplets. There have been numerous attempts to devise non-natural cytosine mimics that provide two appropriate hydrogen bonds to the Hoogsteen face of the guanine. The C-nucleoside analogue 2-aminopyridine has a pK_a of 6.86 compared to that of 4.3 for deoxycytidine itself, and will thus be largely protonated at neutral pH, so ensuring that two hydrogen bonds are likely to be formed (Fig. 4.18). This has been borne out by the formation of stable triplex formation at pH 7 with 2-aminopyridine-containing third strands.[49,50] As yet none of the many cytosine mimics have progressed beyond the experimental stage to become generally available, in large part because of economic considerations.

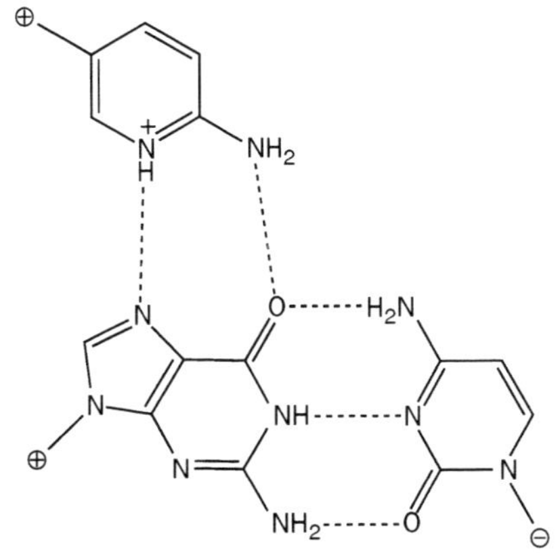

Fig. 4.18 The triplet formed by 2-aminopyridine with a G•C base pair.[49,50]

The relative weakness of the third strand association in triple helices can be enhanced by triplex-selective ligands, which are typically molecules that have the ability to intercalate between base triplets, in a manner analogous to conventional duplex intercalation (see Chapter 5). Molecules such as acridines can be covalently attached to either the 5′ or 3′ end of the third strand oligonucleotide, and rational design considerations have enabled the optimization of a variety of ligands to be undertaken.[51,52] Attachment of a ligand that can covalently cross-link the target duplex, provides yet greater enhancement of binding. Psoralen has been a well-studied, bifunctional, cross-linking agent that binds at the sequence 5′-TpA, thereby placing a restriction on its general applicability for triplexes. In addition psoralen has a requirement for UV activation in order to form covalent cross-links. An alternative approach is to use backbone-modified oligonucleotides for enhancement of stabilization. The use of N3′-P phosphoramidite-modified oligonucleotides is of especial promise, since they hybridize to duplex sequences with significantly higher affinity than unmodified ones, and they are stable in cellular environments—this backbone modification is resistant to nuclease degradation.[53]

4.2.4 Triplex applications

Much of the interest in the triplex phenomenon arises from the findings that triplex formation using an appropriate oligonucleotide can specifically inhibit transcription of a particular gene,[54] for example, by binding to a promotor region that contains a suitable triplex-forming sequence. This is the 'antigene' approach to artificial gene regulation, and complements attempts to read DNA sequences with purely synthetic molecules targeted against the minor groove (see Chapter 5). Both methods require a target site of *c*. 16–20 bases for uniqueness in the human genome, which can readily be achieved by a triplex-forming oligonucleotide provided that the target site has an uninterrupted run of purines or pyrimidines. Since the stability of a triplex is length-dependent, an even longer sequence is to be preferred. The (unsurprising) relative rarity of perfect triplex-forming sequences has encouraged the studies that have been outlined above, which are aimed at extending the current sequence restrictions on triplex formation. The goal of such studies is triplex recognition of a regulatory or coding sequence for *any* gene that is a suitable therapeutic target, and not solely those with perfect oligopyrimidine or oligo-purine triplex sites.

Both parallel and anti-parallel triplex-forming oligonucleotides have been used in a wide range of *in vitro* and cell-based experiments to demonstrate the viability of the triplex approach, especially its ability to down-regulate endogenous genes such as the oncogene c-*myc*.[55] This goal has recently been conclusively demonstrated against a HIV polypurine tract in cell culture,[56] using N3′-P phosphoramidite-modified oligonucleotides to optimize triplex stability and produce inhibition of transcription elongation. Extension of triplex approach to *in vivo* situations is as yet at an early stage of development,[57] not least because of the difficulties involved in ensuring that oligonucleotides are effectively transported into the cell nucleus.

4.3 Guanine quadruplexes

It has long been known that runs of guanosine nucleosides in oligo- and polynucleotides can readily aggregate together, provided a monovalent cation such as potassium or sodium is present to provide stabilization. Diffraction patterns from fibres of poly(G) have been interpreted as arising from a novel four-stranded helix,[58] with four guanine bases all involved in a planar four-stranded (tetrad) arrangement and G · · · G base pairing (Fig. 4.19). This is termed a guanine (G)-quartet, or tetrad. It involves Hoogsteen hydrogen-bonding between the Watson–Crick face of one guanine base and the major-groove face of another, in a manner analogous to that observed in G•G mismatched duplexes.

Guanine-rich sequences occur as overhangs on the 3′ ends of eukaryotic chromosomes (known as telomeres), with repeats of guanines and adenines/thymines, typically of the type G_nT_n, $G_nT_nG_n$, G_nA_n or (TTAGGG)$_n$. Human telomeres in somatic cells are typically 8–10 kilobases long, with the terminal 100–200 bases at the 3′ end being single-stranded. Simple sequences of this type can fold back and then dimerize to form hairpin

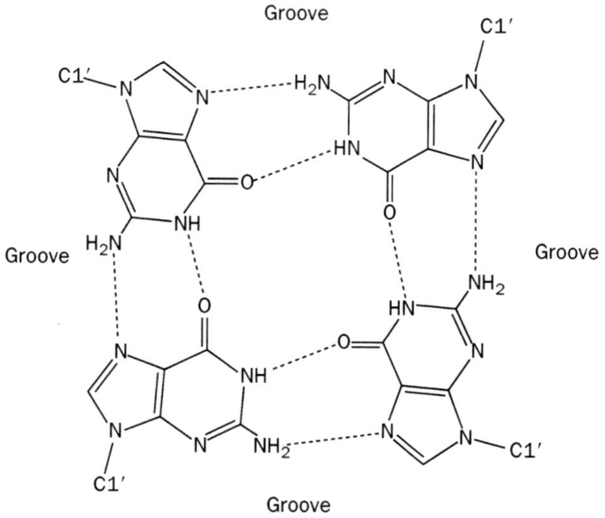

Fig. 4.19 The arrangement of hydrogen bonds in the guanine quartet.

loops that can be stabilized[59] by the formation of the guanine quartets outlined above. Such loops will contain unpaired thymines and are, therefore, potentially relevant models for folded telomeres. These G-quartet-containing structures are termed quadruplexes.[60] The fold-back type outlined above would have anti-parallel strands and an alternating pattern of *syn* and *anti* nucleosides within any one level of quartet although parallel, all-*anti* arrangements are equally possible (see below). The existence of quadruplex structures *in vivo* is still unproven and remains controversial.

There are a number of ways of forming such a structure for any given telomeric sequence. They can be formed by four separate strands associating together in an intermolecular manner (Fig. 4.20), or by a single strand folding back on itself in an intramolecular structure (Fig. 4.21). These two differ in the orientations of the phosphodiester chain within the various strands of the complex. Whether the strands are parallel or anti-parallel to each other depends, in part, on the nature and length of the sequence, as well as the counter-ion used.

The sequence d(GGGGTTTTGGGG), from the telomere of the *Oxytricha* organism, has been extensively studied by structural methods. The general arrangement is of a four-stranded, folded quadruplex with two strands associating together as an intramolecular hairpin structure. There are four stacked hydrogen-bonded G-quartets (Fig. 4.19), sandwiched on top of each other at a separation of 3.4 Å. An early crystallographic analysis of the potassium complex[61] suggested that the quadruplex has adjacent strands anti-parallel to each other with the loops joining adjacent rather than diagonal strands (Fig. 4.22a). There is alternation of *syn* and *anti* glycosidic angles along each chain. On the other hand, NMR studies[62,63] on the structures of this sequence with all three sodium, potassium and ammonium ion forms, do not observe the same anti-parallel arrangement. Instead, the adjacent strands are alternately parallel and anti-parallel (Fig. 4.22b). There are still alternating *syn* and *anti* glycosidic conformations along any one strand, but the arrangement of strands results in a G-quartet arrangement with a *syn-syn–anti-anti* pattern

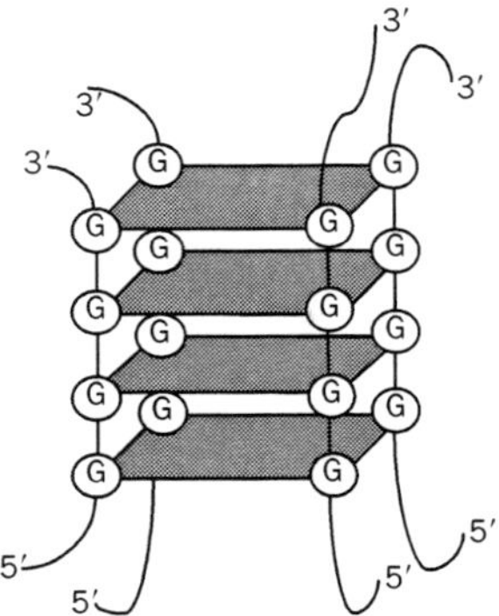

Fig. 4.20 Strand polarities in an all-parallel guanine-quadruplex with four G-quartets. Figures drawn by Martin Read, Chester Beatty Laboratories.

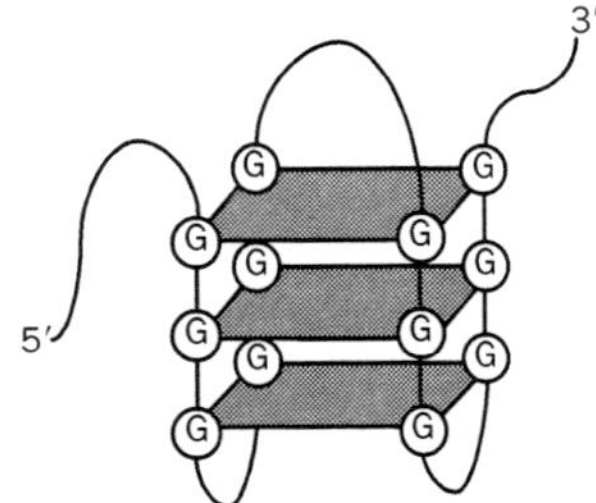

Fig. 4.21 Schematic view of the folding of DNA strands in an intramolecular quadruplex with three G-quartets.

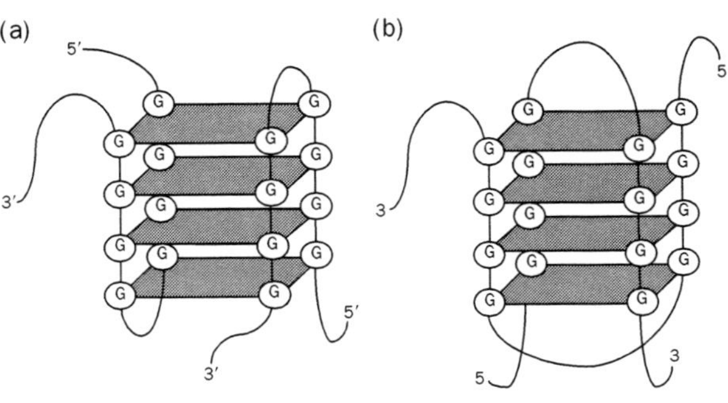

Fig. 4.22 Two possible folding arrangements for a hairpin dimer quadruplex formed from two DNA strands.

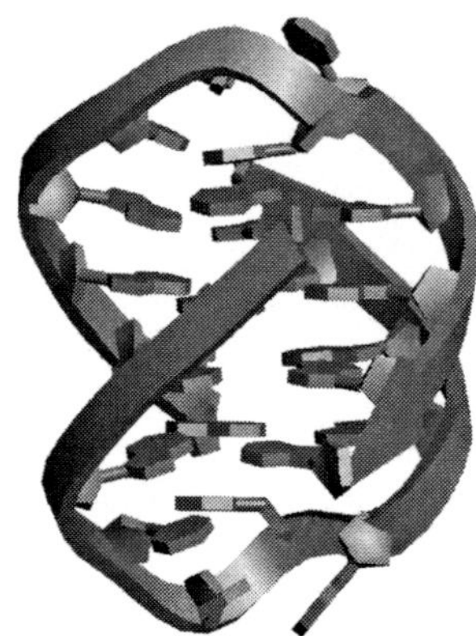

Fig. 4.23 The crystal structure[65] of the hairpin quadruplex formed by two strands of d(GGGGTTTGGGG), with the same fold as in Fig. 4.22(b).

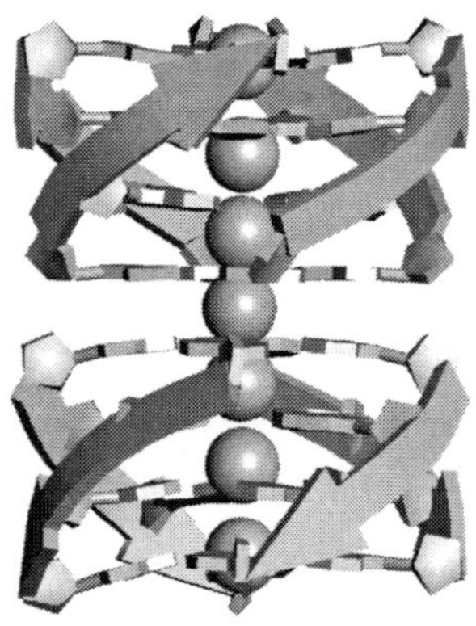

Fig. 4.24 Two parallel d(TGGGGT) quadruplexes, as found stacked in the crystal structure,[69] showing the sodium ions as shaded spheres.

of glycosidic angles and loops joining diagonal strands. This overall fold is retained with all of these counter-ions. More recent crystal structure analyses of this sequence have been reported, one with sodium ions, and co-crystallized within the structure of a protein-single-stranded telomere complex,[64] and the other as the native structure in the presence of potassium ions.[65] Both show the same fold and strand polarities as in the NMR structures.[62,63] There are some differences between NMR and crystal structures that are mostly concerned with the conformations of thymines in the loops. In the latter,[64,65] the thymines are highly ordered and there is extensive stacking between them (Fig. 4.23).

The counter-ions play a key role in maintaining the integrity of quadruplex structures. They do this by forming ionic interactions with O6 atoms of the guanines. At one extreme an ion can be positioned symmetrically out of the plane of four guanines in a quartet. Sometimes an ion is situated symmetrically between two consecutive quartets so that it is eight co-ordinated, to O6 atoms from both quartets. This arrangement is favoured by potassium and ammonium ions. The smaller sodium ion is able to coordinate all four O6 atoms in a quartet while being in the plane of all four guanines, although it can also be accommodated midway between quartets. A continuum between both extreme positions can be seen in several of the crystal and NMR structures,[63,64] with the formation of a continuous channel of positive charge along the quadruplex. There is some evidence that sequences related to d(GGGGTTTTGGGG), but with asymmetric loops, may be more prone to significant counter-ion conformational flexibility.

A number of other sequences have been studied by NMR methods; some, such as d(TGGGGT)[66] and d(TTGGGG),[67] which form parallel-stranded quadruplex structures with all-*anti* glycosidic angles. Others such as d(GGTTTTCGG),[68] form anti-parallel arrangements with alternating *syn–anti* glycosidic bonds. The parallel d(TGGGGT) quadruplex structure, stabilized by sodium ions, has been studied in detail by X-ray crystallography.[69] The structure, which contains four independent quadruplexes in the crystallographic asymmetric unit, has been solved to exceptionally high resolution (0.95 Å), enabling many details of the structure to be precisely defined (Table 4.5). Two quartets are stacked together in the crystal, with a channel of seven sodium ions in the centre (Fig. 4.24). There are a range of positions for these ions (and hence of ion–ion distances) with respect to the quartet planes, although there is a clear pattern of increasing coplanarity with quartets on moving towards the ends of the tetraplex dimer. Most of the sugar rings are in the normal B-DNA C3′-*endo* conformation. However, the distribution of values for the backbone torsion angles δ and ε are outside the normal duplex ranges, suggesting that

Table 4.5 Crystal structures of nucleic acid quadruplexes

Sequence	NDB code no.
d(GGGGTTTTGGGG) + K$^+$	UD0013
d(GGGGTTTTGGGG) + Na$^+$	PD0218
d(TGGGGT)	UDF062
d(GCATGCT)	UDG028

the quadruplex backbone is strained, possibly in order for the hydrogen bonding of the quartets to occur.

4.3.1 **Complex quartet structures**

A more complex arrangement for a quadruplex fold has been found in the intramolecular fold-back quadruplex (Figs 4.21 and 4.25) formed by the sequence d[AGGG-(TTAGGG)$_3$], that is, almost four repeats of the human telomeric sequence. The NMR structure[70] with sodium ions shows that each strand is flanked by one parallel and one anti-parallel neighbours. There is a core of three guanine quartets flanked by two lateral and one diagonal TTA loops. The asymmetry of the loops results in significant differences between the widths of the four grooves (Fig. 4.25). For the NMR structure with the lowest rmsd, two grooves have intermediate widths of 11.5–13.5 Å, and are flanked by a narrow (<10 Å) and a wide groove (>15 Å). These differences may be of significance for the recognition of quadruplexes by telomere-binding proteins.[71]

The realization that quadruplex structures could be formed by bases other than solely guanine has lead to the determination of several remarkable folded DNA structures.[72] Such structures may be formed by G-rich sequences when used as aptamers for highly specific binding to protein targets. The short sequence d(GCATGCT) forms a loop arrangement which associates into an anti-parallel dimer (Fig. 4.26) with two CGCG quartets.[73] Even more remarkably, adenines have been found[74] to associate with a conventional G-quartet in the structure of the sequence d(GGAGGAG) to form A•G•G•G•G•A hydrogen-bonded hexads stacked between two G-quartets (Fig. 4.27). Although many more possible arrangements

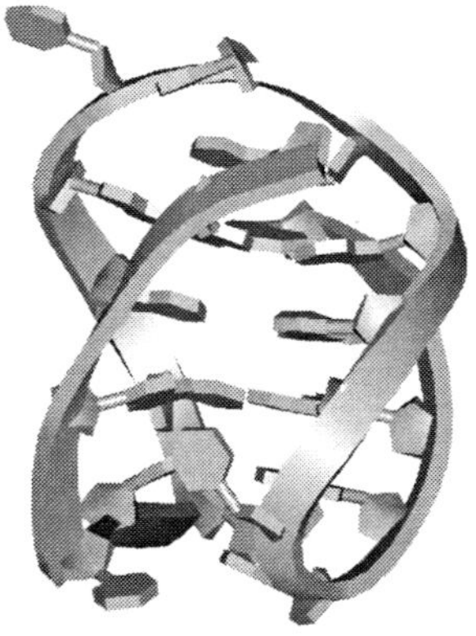

Fig. 4.25 The NMR structure[70] of the sodium form of the intramolecular quadruplex formed by d(AGGG(TTAGGG)$_3$).

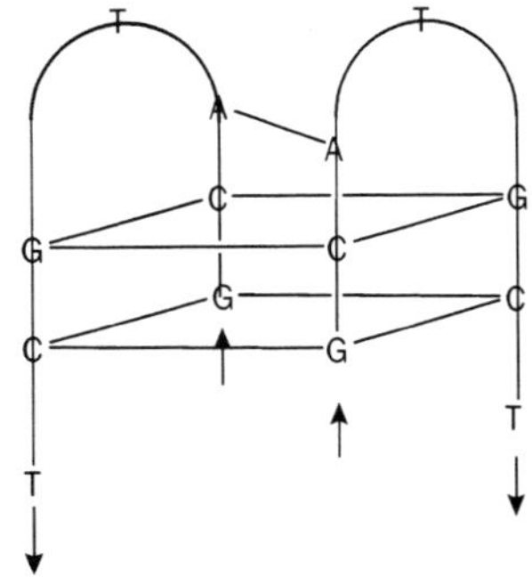

Fig. 4.26 Schematic representation of the quadruplex formed by the sequence d(GCATGCT), as found in its crystal structure.[73]

for these types of mixed G-rich sequences remain to be thoroughly explored, it is now possible to design appropriate sequences that will incorporate features such as quartets and triplets in defined relationships to each other.[75,76]

Molecular dynamics simulations of G-quadruplexes over multi-nanosecond time-scales have shown that the G-quartet arrangement is remarkably stable, especially in the parallel

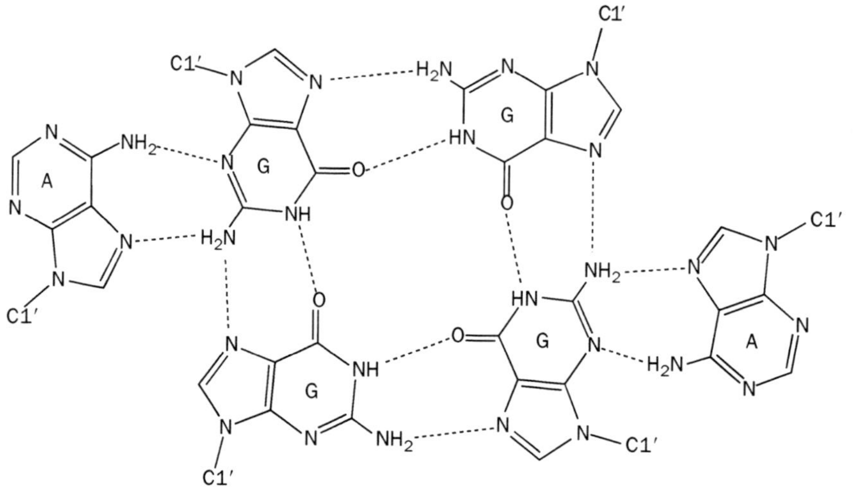

Fig. 4.27 The hydrogen bonding postulated for the A•G•G•G•G•A hexad.[74]

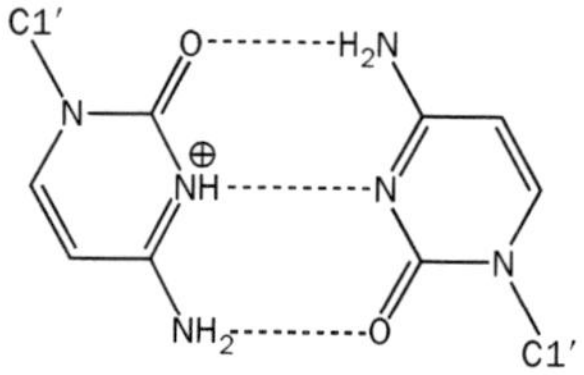

Fig. 4.28 The hydrogen bonding arrangement in a C · · · C+ base pair.

d(TGGGGT) quadruplex.[77,78] Simulations have included sodium and potassium counter-ions as well as water molecules, and so have been able to ascertain the roles played by ions in both conventional hairpin dimers with standard G-quartets,[79] and in quadruplexes with mixed GCGC quartets.[80] The latter appear to require just two cations for a structure comprising two G-quartets and two CGCG ones, with subtle differences in the hydrogen bonding arrangement being dependent on the nature of the cation.

Some general patterns of the relationships between strand polarity and base pairing, have emerged from studies on triplexes and quadruplexes, as well as on Watson–Crick, left-handed and mispaired duplexes.[81,82] For example, the pattern of glycosidic angles for a base pair is a consequence of the relationship between strand directions in quadruplexes, with *syn* angles being produced by anti-parallel strands and *anti* angles by parallel ones. This has led to a general classification scheme for nucleic acid structural types,[82] based on these polarity and base pairing criteria. Prediction of new structural types emerges naturally from the scheme. The variety of DNA structures that have been established over the past few years and described in this Chapter suggests that many more remain to be experimentally determined.

4.3.2 **The i-motif**

Diffraction studies on poly (C) fibres produced at acidic pH were interpreted[83] in terms of a parallel-stranded duplex with C•C+ mismatch base pairs (Fig. 4.28). As with guanine-quadruplexes, this type of arrangement was largely forgotten until recently. An NMR study on the cytosine-rich oligonucleotide d(TCCCC) showed that this sequence formed, not a duplex, but instead a novel four-stranded quadruplex,[84] which was termed

Table 4.6 Crystal structures of nucleic acid i-motifs

Sequence	NDB code no.
d(CCCC)	UDD024
d(CCCT)	UDD023
d(ACCCT)	UD0003
d(TAACCC)	UDF027
d(CCCAAT)	UDF043
d(AACCCC)	UDF054

the i-motif (Table 4.6). The basic unit is the C•C$^+$ mismatch base pair, with one of the two cytosine protonated at the N3 position such that three hydrogen bonds are formed.

The arrangement of the i-motif bears some resemblance to the original fibre diffraction model in that the two strands connected by the C•C$^+$ base pairs are in a parallel orientation. Two of these duplexes are intercalated into each other, which results in sets of consecutive C•C$^+$ base pairs oriented approximately at right angles to each other (Fig. 4.29). The DNA strand complementary to a guanine-rich, double-stranded centromeric, or telomeric sequence is by definition cytosine-rich. A number of crystal structures of such sequences have been reported, all of which show this structural motif.[85–89] These show that the grooves in the i-motif structures are highly asymmetric. There are two broad, shallow major grooves, of width typically 16 Å. Each is flanked by a very narrow minor groove (Fig. 4.30), of width around 6 Å. The helical twist between consecutive C•C$^+$ base pairs in either duplex tends to be small, between 12° and 20°.

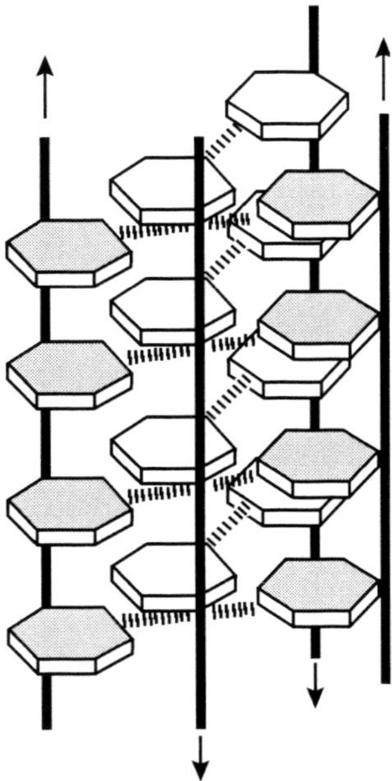

Fig. 4.29 Schematic view of the arrangement of base pairs in an i-motif four-stranded structure. The base pairs shown shaded form one double helix, which is intercalated into that formed by the unshaded base pairs.

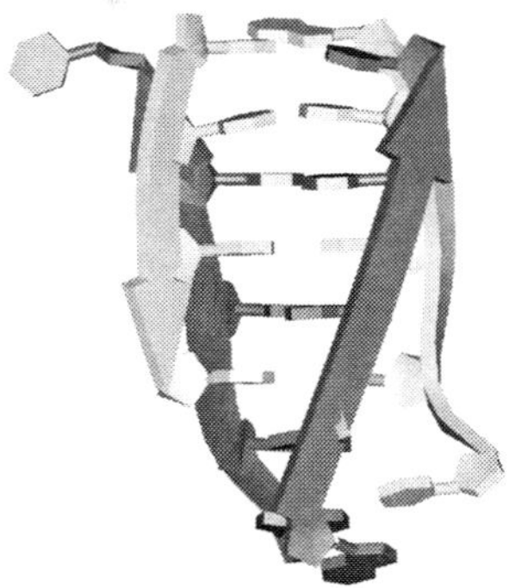

Fig. 4.30 Structure of the i-motif sequence d(), as found in the crystal structure.[89]

4.4 **DNA junctions**

The process of homologous genetic recombination involves the crossing-over of strands from two DNA sequences. The central intermediate in this process is believed to be a four-way branched junction-type structure (Fig. 4.31), termed a Holliday junction. It consists of four strands, each participating in four anti-parallel base-paired arms. A wide variety of biophysical techniques, such as gel electrophoresis and fluorescence energy transfer have been used in attempts to define the three-dimensional structure of these junctions.[90] Data from these studies,[91] together with molecular modelling,[92] resulted in the proposal of the X-structure, with two pairs of almost parallel helices and a region in the centre where the DNA strands from the two starting duplexes cross over.

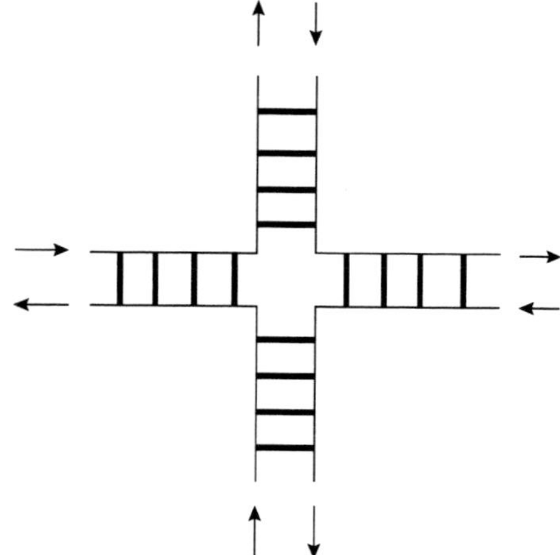

Fig. 4.31 The organization of a four-way Holliday junction. The directionality of each of the four strands is shown by arrows.

Table 4.7 Crystal structures of nucleic acid junctions

Sequence	NDB code no.
d(CCGGTACCGG)	UD0008
d(CCGGGACCGG)	UD0006
10–23 DNA enzyme + RNA	UH0001
10–23 DNA enzyme + RNA	UH0002

There have been numerous attempts to crystallize junction-type DNA structures,[90] mostly involving long sequences (Table 4.7). Success was finally achieved independently in two laboratories, very surprisingly with two short, closely similar decamer sequences. The first, d(CCGGGACCGG), was crystallized[93] with the goal of studying contiguous G•A mismatches in duplex DNA. The crystal structures of the fully inverted-repeat sequences d(CCGGTACCGG) and d(CCGCTAGCGG) shows a closely similar arrangement,[94] albeit without any possible perturbing effects of the G•A mismatch. All three structures consist of two pairs of arms of the junction stacked on each other to form two continuous antiparallel duplexes that have totally normal linear B-DNA conformations (Figs 4.32 and 4.33). All bases are in base paired arrangements in the duplex regions. The junction itself is formed by a few changes in backbone conformations, chiefly in the two cytosine-7 nucleotides on the two inner-facing strands. The two arms are right-hand twisted with an angle between them of ~40°, in accord with the stacked X-junction model—it is remarkable how close the crystal structure matches the conclusions from the earlier biophysical and modelling studies. The 5′-ACC sequence is at the core of the junction and may be important for its stabilization. The closeness of four phosphate groups at the junction requires compensating cationic charge, and a sodium ion has been located in a buried position within the junction, bridging two adenosine-6 phosphate oxygen atoms and situated on the local two-fold axis of the structure.

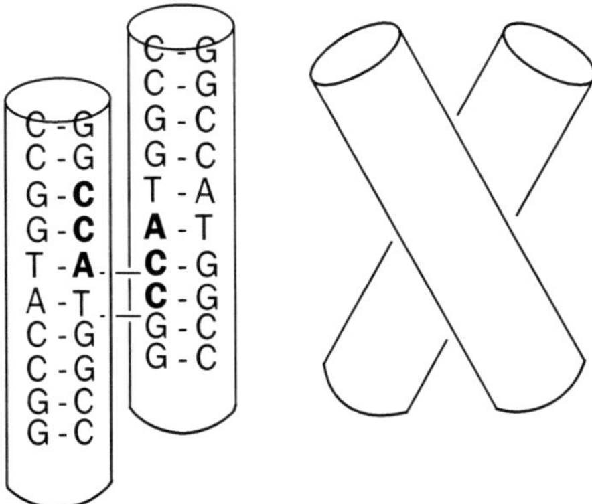

Fig. 4.32 Schematic views of a Holliday junction structure, showing, on the left, the two helical arms and their connection points via the two central cross-over sequences. The relative angle of the two helical arms is shown on the right.

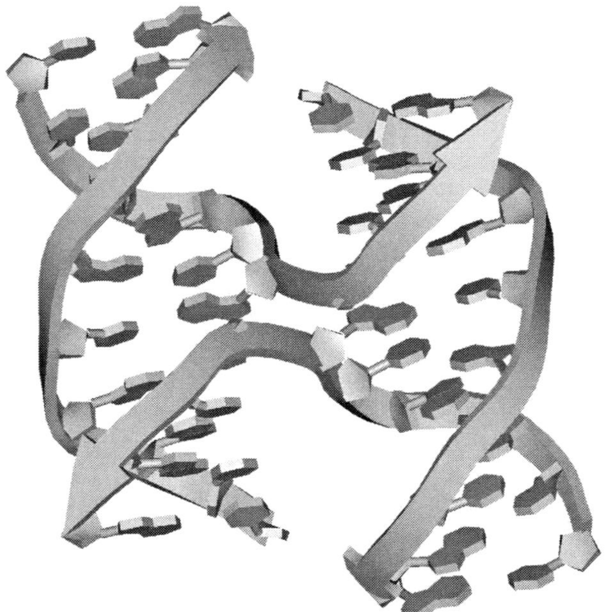

Fig. 4.33 Structure of the Holliday junction formed by four decamer strands.[94]

More complex junction structures have been found in the crystal structures of 'DNA enzymes'. These analyses follow the discovery, using *in vitro* selection procedures, that certain DNA sequences can act as enzymes, for example, being able to cleave phosphodiester bonds.[95] The '10–23' motif, which catalyzes the cleavage of RNA sequences, has been co-crystallized[96] with an inactivated RNA substrate. The resulting crystal structure consists of a dimer of the DNA : RNA complex in which there are two DNA and two RNA strands. These form five double-helical regions (Fig. 4.34). A four-way junction is produced in this structure by the confluence of the hexanucleotide loop with three stem regions. As in the Holliday junction structures, all bases are base paired and stacked. All of these junction structures have an overall X-form, and there are few differences

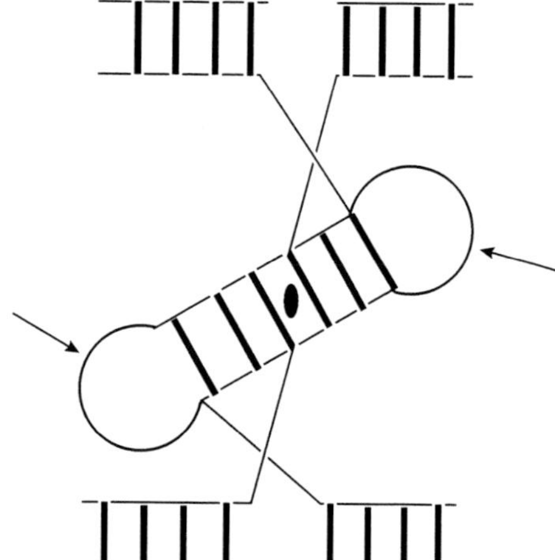

Fig. 4.34 Schematic view of the connectivity of the helices and loop regions in the crystal structure of the 10–23 DNA enzyme.[96]

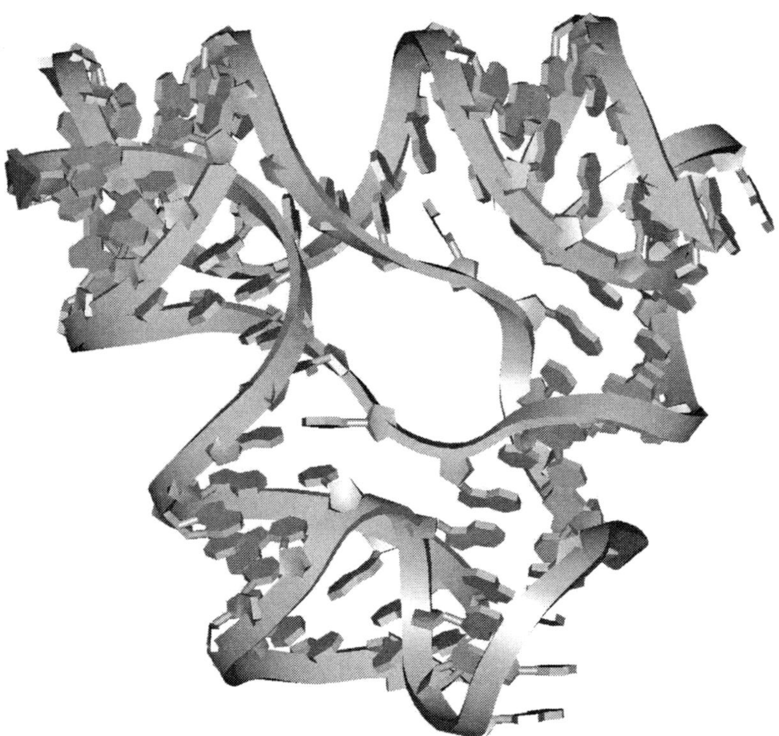

Fig. 4.35 Crystal structure of the 108-mer DNA enzyme.[96]

between them, apart from the A-form nature of two stems in the 10–23 junction (which arises from the involvement of RNA). By contrast, a 108-nucleotide DNA–RNA complex[97] adopts an X-junction structure but with a left-handed orientation of the arms (Fig. 4.35).

References

1   Patel, D. J. (1992). *Current Opinion in Structural Biology*, **2**, 345.

2   Brown, T., Hunter, W. N., Kneale, G., and Kennard, O. (1986). *Proceedings of the National Academy of Sciences USA*, **83**, 2402.

3   Leonard, G. A., Booth, E. D., and Brown, T. (1990). *Nucleic Acids Research*, **18**, 5617.

4   Webster, G. D., Sanderson, M. R., Skelly, J. V., Neidle, S., Swann, P. F., Li, B. F., and Tickle, I. J. (1990). *Proceedings of the National Academy of Sciences USA*, **87**, 6693.

5   Privé, G. G., Heinemann, U., Chandrasegaran, S., Kan, L.-S., Kopka, M. L., and Dickerson, R. E. (1987). *Science*, **238**, 498.

6   Skelly, J. V., Edwards, K. J., Jenkins, T. C., and Neidle, S. (1993). *Proceedings of the National Academy of Sciences USA*, **89**, 804.

7   Chou, S.-H., Cheng, J.-W., and Reid, B. R. (1992). *Journal of Molecular Biology*, **228**, 138.

8   Gao, X. and Patel, D. J. (1988). *Journal of the American Chemical Society*, **110**, 5178.

9   Lane, A. N., Jenkins, T. C., Brown, D. J. S., and Brown, T. (1991). *Biochemical Journal*, **279**, 269.

10 Li, Y., Zon, G., and Wilson, W. D. (1991). *Proceedings of the National Academy of Sciences USA*, **88**, 26.

11 Li, Y., Zon, G., and Wilson, W. D. (1991). *Biochemistry*, **30**, 7566.

12 Gao, Y.-G., Robinson, H., Sanishvili, R., Joachimiak, A., and Wang, A. H.-J. (1999). *Science*, **38**, 16452.

13 Green, K. L., Jones, R. L., Li, Y., Robinson, H., Wang, A. H.-J., and Wilson, W. D. (1994). *Biochemistry*, **33**, 1053.

14 Chou, S.-H., Zhu, L., and Reid, B. R. (1997). *Journal of Molecular Biology*, **267**, 1055.

15 Kalnik, M. W., Li, B. F. L., Swann, P. F., and Patel, D. J. (1989). *Biochemistry*, **28**, 6182.

16 Ginell, S. L., Kuzmich, S., Jones, R. A., and Berman, H. M. (1990). *Biochemistry*, **29**, 10461.

17 Ginell, S. L., Vojtechovsky, J., Gaffney, B., Jones, R. A., and Berman, H. M. (1994). *Biochemistry*, **33**, 3487.

18 Leonard, G. A., Thomson, J., Watson, W. P., and Brown, T. (1990). *Proceedings of the National Academy of Sciences USA*, **87**, 9573.

19 Vojtechovsky, J., Eaton, M. D., Gaffney, B., Jones, R. A., and Berman, H. M. (1995). *Biochemistry*, **34**, 16632.

20 Chatake, T., Ono, A., Ueno, Y., Matsuda, A., and Takénaka, A. (1999). *Journal of Molecular Biology*, **294**, 1215.

21 Chatake, T., Hikima, T., Ono, A., Ueno, Y., Matsuda, A., and Takénaka, A. (1999). *Journal of Molecular Biology*, **294**, 1223.

22 Lipscomb, L. A., Peek, M. E., Morningstar, M. L., Verghis, S. M., Miller, E. M., Rich, A., *et al.* (1995). *Proceedings of the National Academy of Sciences USA*, **92**, 719.

23 Felsenfeld, G., Davies, D. R., and Rich, A. (1957). *Journal of the American Chemical Society*, **79**, 2023.

24 Arnott, S., Bond, P. J., Selsing, E., and Smith, P. J. C. (1976). *Nucleic Acids Research*, **11**, 4141.

25 Strobel, S. A. and Dervan, P. B. (1990). *Science*, **249**, 73.

26 de los Santos, C., Rosen, M., and Patel, D. J. (1989). *Biochemistry*, **28**, 7282.

27 Radhakrishnan, I., Patel, D. J., Veal, J. M., and Gao, X. L. (1992). *Journal of the American Chemical Society*, **114**, 6913.

28 Macaya, R., Wang, E., Schultze, P., Sklenár, V., and Feigon, J. (1992). *Journal of Molecular Biology*, **225**, 755.

29 Radhakrishnan, I and Patel, D. J. (1994). *Biochemistry*, **33**, 1405.

30 Gotfredsen, C. H., Schultze, P., and Feigon, J. (1998). *Journal of the American Chemical Society*, **120**, 4281.

31 Cheng, Y.-K. and Pettitt, B. M. (1992). *Journal of the American Chemical Society*, **114**, 4465.

32 Laughton, C. A. and Neidle, S. (1992). *Journal of Molecular Biology*, **223**, 519.

33 Shields, G. C., Laughton, C. A., and Orozco, M. (1997). *Journal of the American Chemical Society*, **119**, 7463.

34 Liu, K., Sasisekharan, V., Miles, H. T., and Raghunathan, G. (1996). *Biopolymers*, **39**, 573.

35 Van Meervelt, L., Vlieghe, D., Dautant, A., Gallois, B., Précigoux, G., and Kennard, O. (1995). *Nature*, **374**, 742.

36 Betts, L., Joisey, J. A., Veal, J. M., and Jordan, S. R. (1995). *Science*, **270**, 1838.

37 Nielsen, P. E., Egholm, M., Berg, R. H., and Buchardt, O. (1991). *Science*, **254**, 1497.

38 Uhlmann, E. (1998). *Biological Chemistry*, **379**, 1045.

39   Rhee, S., Han, Z., Liu, K., Miles, H. T., and Davies, D. R. (1999). *Biochemistry*, **38**, 16810.

40   Beal, P. A. and Dervan, P. B. (1991). *Science*, **251**, 1360.

41   Radhakrishnan, I., de los Santos, C., and Patel, D. J. (1991). *Journal of Molecular Biology*, **221**, 1403.

42   Laughton, C. A. and Neidle, S. (1992). *Nucleic Acids Research*, **20**, 6535.

43   Gowers, D. M. and Fox, K. R (1999). *Nucleic Acids Research*, **27**, 1569.

44   Griffin, L. C. and Dervan, P. B. (1989). *Science*, **245**, 967.

45   Mergny, J.-L., Sun, J.-S., Rougée, M., Montenay-Garestier, T., Barcelo, F., Chomilier, J., and Hélène, C. (1991). *Biochemistry*, **30**, 9791.

46   Kiessling, L. L., Griffin, L. C., and Dervan, P. B. (1992). *Biochemistry*, **31**, 2829.

47   Griffin, L. C., Kiessling, L. L., Beal, P. A., Gillespie, P., and Dervan, P. B. (1992). *Journal of the American Chemical Society*, **114**, 7976.

48   Bijapur, J., Keppler, M. D., Bergqvist, S., Brown, T., and Fox, K. R. (1999). *Nucleic Acids Research*, **27**, 1802.

49   Bates, P. J., Laughton, C. A., Jenkins, T. C., Capaldi, D. C., Roselt, P. D., Reese, C. B., and Neidle, S. (1996). *Nucleic Acids Research*, **24**, 4176.

50   Hildebrand, S., Blaser, A., Parel, S., and Leumann, C. J. (1997). *Journal of the American Chemical Society*, **119**, 5499.

51   Escudé, C., Nguyen, C. H., Kukreti, S., Janin, Y., Sun, J.-S., Bisagni, E., *et al.* (1998). *Proceedings of the National Academy of Sciences USA*, **95**, 3591.

52   Keppler, M. D., Neidle, S., and Fox, K. R. (2001). *Nucleic Acids Research*, **29**, 1953.

53   Escudé, C., Giovannangeli, C., Sun, J.-S., Lloyd, D. H., Chen, J. K., Gryaznov, S., *et al.* (1996). *Proceedings of the National Academy of Sciences USA*, **93**, 4365.

54   Grigoriev, M., Praseuth, D., Guieysse, A. L., Robin, P., Thoung, N. T., Hélène, C., and Harel-Bellan, A. (1993). *Proceedings of the National Academy of Sciences USA*, **90**, 3501.

55   Catapano, C. V., McGuffie, E. M., Pacheco, D., and Carbone, G. M. R. (2000). *Biochemistry*, **39**, 5126.

56   Faria, M., Wood, C. D., Perrouault, L., Nelson, J. S., Winter, A., White, M. R. H., *et al.* (2000). *Proceedings of the National Academy of Sciences USA*, **97**, 3862.

57   Giovannangeli, C. and Hélène, C. (2000). *Nature Biotechnology*, **18**, 1245.

58   Arnott, S., Chandrasekaran, R., and Marttila, C. M. (1974). *Biochemical Journal*, **141**, 537.

59   Sundquist, W. I. and Klug, A. (1989). *Nature*, **342**, 825.

60   Gilbert, D. E. and Feigon, J. (1999). *Current Opinion in Structural Biology*, **9**, 305.

61   Kang, C., Zhang, X., Ratliff, R., Moyzis, R., and Rich, A. (1992). *Nature*, **356**, 126.

62   Smith, F. W. and Feigon, J. (1992). *Nature*, **356**, 164.

63   Schultze, P., Hud, N. V., Smith, F. W., and Feigon, J. (1999). *Nucleic Acids Research*, **27**, 3018.

64   Horvath, M. P. and Schultz, S. C. (2001). *Journal of Molecular Biology*, **310**, 367.

65   Haider, S., Parkinson, G., and Neidle, S. (2001). To be published.

66   Aboul-ela, F., Murchie, A. I. H., and Lilley, D. M. J. (1992). *Nature*, **360**, 280.

67   Phillips, K., Dauter, Z., Murchie, A. I. H., Lilley, D. M. J., and Luisi, B. J. (1997). *Journal of Molecular Biology*, **273**, 171.

68   Wang, Y. and Patel, D. J. (1992). *Biochemistry*, **31**, 8112.

69   Wang, Y., de los Santos, C., Gao, X., Greene, K., Live, D., and Patel, D. J. (1991). *Journal of Molecular Biology*, **222**, 819.

70   Wang, Y. and Patel, D. J. (1993). *Structure*, **1**, 263.

71   Arthanari, H. and Bolton, P. H. (2001). *Chemistry and Biology*, **8**, 221.

72   Patel, D. J., Bouaziz, S., Kettant, A., and Wang, Y. (1999). In *Oxford Handbook of Nucleic Acid Structure* (ed. Neidle, S), pp. 389–453, Oxford.

73   Leonard, G. A., Zhang, S., Peterson, M. R., Harrop, S. J., Helliwell, J. R., Cruse, W. B. T., *et al.* (1995). *Structure*, **3**, 335.

74   Kettani, A., Gorin, A., Majumdar, A., Hermann, T., Skripkin, E., Zhao, H., *et al.* (2000). *Journal of Molecular Biology*, **297**, 627

75   Kettani, A., Basu, G., Gorin, A., Majumdar, A., Hermann, T., Skripkin, E., and Patel, D. J. (2000). *Journal of Molecular Biology*, **301**, 129

76 Kuryavyi, V., Majumdart, A., Shallop, A., Chernichenko, N., Skripkin, E., Jones, R., and Patel, D. J. (2001). *Journal of Molecular Biology*, **310**, 181.

77 Spačková, N., Berger, I., and Šponer, J. (1999). *Journal of the American Chemical Society*, **121**, 5519.

78 Read, M. A. and Neidle, S. (2000). *Biochemistry*, **39**, 13422.

79 Strahan, G. D., Keniry, M. A., and Shafer, R. H. (1998). *Biophysical Journal*, **75**, 968.

80 Špa ková, N., Berger, I., and Šponer, J. (2001). *Journal of the American Chemical Society*, **123**, 3295.

81 Westhof, E. (1992). *Nature*, **358**, 459.

82 Lavery, R., Zakrzewska, K., Sun, J.-S., and Harvey, S. C. (1992). *Nucleic Acids Research*, **20**, 5011.

83 Langridge, R. and Rich, A. (1963). *Nature*, **198**, 725.

84 Gehring, K., Leroy, J.-L., and Guéron, M. (1993). *Nature*, **363**, 561.

85 Kang, C., Berger, I., Lockshin, C., Ratliff, R., Moysis, R., and Rich, A. (1994). *Proceedings of the National Academy of Sciences USA*, **91**, 11636.

86 Kang, C., Berger, I., Lockshin, C., Ratliff, R., Moysis, R., and Rich, A. (1995). *Proceedings of the National Academy of Sciences USA*, **92**, 3874.

87 Berger, I., Kang, C., Fredian, A., Lockshin, C., Ratliff, R., Moysis, R., and Rich, A. (1995). *Nature Structural Biology*, **2**, 416.

88 Cai, L., Chen, L., Raghavan, S., Ratliff, R., Moysis, R., and Rich, A. (1998). *Nucleic Acids Research*, **26**, 4696.

89 Weil, J., Min, T., Yang, C., Wang, S., Sutherland, C., Sinha, N., and Kang, C. (1999). *Acta Crystallographica*, **D55**, 422.

90 Lilley, D. M. J. and Norman, D. G. (1999). *Nature Structural Biology*, **6**, 897.

91 Duckett, D. R., Murchie, A. I. H., Giraud-Panis, M.-J. E., Pöhler, J. R., and Lilley, D. M. J. (1995). *Philosophical Transactions of the Royal Society*, **B347**, 27.

92 Von Kitzing, E., Lilley, D. M. J., and Diekmann, S. (1990). *Nucleic Acids Research*, **18**, 2671.

93 Ortiz-Lombardia, M., González, A., Eritja, R., Aymami, J., Azolin, F., and Coll, M. (1999). *Nature Structural Biology*, **6**, 913.

94 Eichman, B. F., Vargason, J. M., Mooers, B. H. M., and Ho, P. S. (2000). *Proceedings of the National Academy of Sciences USA*, **97**, 3971.

95 Breaker, R. R. and Joyce, G. F. (1994). *Chemistry and Biology*, **1**, 223.

96 Nowakowski, J., Shim, P. J., Prasad, G. S., Stout, C. D., and Joyce, G. F. (1999). *Nature Structural Biology*, **6**, 151.

97 Nowakowski, J., Shim, P. J., Stout, C. D., and Joyce, G. F. (1999). *Journal of Molecular Biology*, **300**, 93.

Further reading

Mispairing

Brown, T. and Kennard, O. (1992). *Current Opinion in Structural Biology*, **2**, 354.

Hunter, W. N. (1992). *Methods in Enzymology*, **211**, 221.

Modrich, P. (1987). *Annual Reviews in Biochemistry*, **56**, 435.

Modrich, P. (1991). *Annual Reviews in Genetics*, **25**, 229.

Modrich, P. and Lahue, R (1996). *Annual Reviews in Biochemistry*, **65**, 101.

Singer, B. and Grunberger, D. (1983). *Molecular Biology of Mutagens and Carcinogens*, Plenum Press, New York.

Triplexes

Fox, K. R. (2000). *Current Medicinal Chemistry*, **7**, 17.

Moser, H. and Dervan, P. B. (1987). *Science*, **238**, 645.

Neidle, S. (1997). *Anti-Cancer Drug Design*, **12**, 433.

Soyfer, V. N. and Potaman, V. N. (1996). *Triple Helical Nucleic Acids*, Springer-Verlag, New York.

Sun, J. S., Garestier, T., and Hélène, C. (1996). *Current Opinion in Structural Biology*, **6**, 327.

Thuoung, N. T. and Hélène, C. (1993). *Angewante Chemie, International Edition*, **32**, 666.

Telomeres and quadruplexes

Blackburn, E. H. (2000). *Nature*, **408**, 53.

Cech, T. (2000). *Angewante Chemie, International Edition*, **39**, 34.

Gilbert, D. E. and Feigon, J. (1999). *Current Opinion in Structural Biology*, **9**, 305.

Kerwin, S. M. (2000). *Current Pharmaceutical Design*, **6**, 441.

Krauskopf, A. and Blackburn, E. H. (2000). *Annual Reviews of Genetics*, **34**, 331.

Williamson, J. R. (1994). *Annual Reviews of Biophysics*, **23**, 703.

DNA junctions

Lilley, D. M. J. (2000). *Quarterly Reviews in Biophysics*, **33**, 109.

Ho, P. S. and Eichman, B. F. (2001). *Current Opinion in Structural Biology*, **11**, 302.

5

Principles of small molecule–DNA recognition

5.1 Introduction

The recognition of a DNA molecule can, on the one hand, be general for any individual nucleotide or sequence. At the other extreme, it can be highly specific, solely recognizing a defined sequence within a genome. This can be performed very effectively by DNA itself, as has been seen in Chapter 4, with DNA–DNA recognition being governed in large part by base–base hydrogen-bonding. A single strand of oligonucleotide can recognize

- another single strand to form a duplex
- a duplex to form a triplex
- or even three other strands to form a quadruplex.

In each case there is an increasingly stringent sequence requirement. All this occurs by exploitation of the hydrogen bonding ability of DNA bases over and above Watson–Crick hydrogen bonding. It is also the manner in which direct readout of DNA sequence information can be achieved by small molecules and proteins.

The task of DNA sequence recognition has as its prerequisite a knowledge of the frequency of occurrence of a given DNA sequence within a defined length of DNA. This can be directly obtained from a DNA databank if the entire target sequence is known. This can now be done for the human genome, and for a rapidly increasing number of other genomes in many species. There have been several estimates of frequency of occurrence, on the basis of assumptions of randomness for the distribution of the four bases, and of sequences generally. It is then possible to calculate the frequency of occurrence for a given size of a DNA recognition site. The reciprocal of the former is the number of unique sites for a given site size. The data in Table 5.1 suggests that the human genome, with a size of 3.2 billion nucleotides, requires a recognition site of 16–18 nucleotides in order to be reasonably certain of uniqueness, that is, if a single sequence within a gene of therapeutic interest is to be targeted. In fact there appear to be only *c.* 30 000 genes coded by the entire genome, corresponding to about 1–2 per cent of the human genome.[1] Thus the sequence recognition problem is in reality just slightly less daunting, and the size of site to be recognized uniquely may be to the lower end of the

Table 5.1 The statistics of DNA sequence selectivity, based on a random distribution of nucleotides

Site size	Frequency *n*, as 1 : *n*
2	10
3	32
4	136
5	512
6	2080
7	8196
10	524 800
12	8 390 656
15	536 870 912
16	2 147 516 416

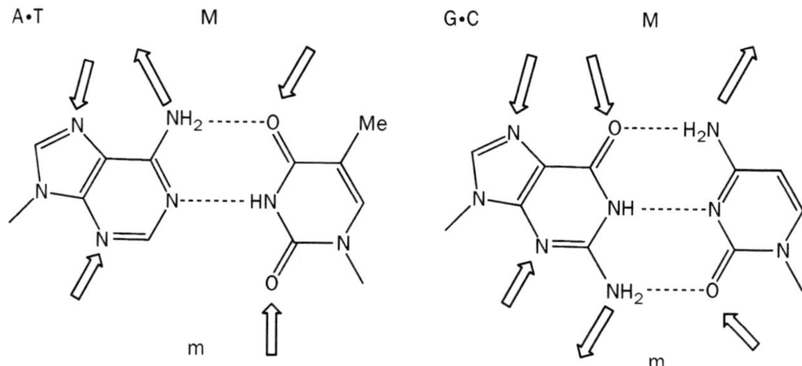

Fig. 5.1 The pattern of hydrogen-bond donors and acceptors in A•T and G•C base pairs. The arrows indicate the directionality of the hydrogen bonds to and from the base edges. M and m indicate the major and minor grooves respectively.

16–18 base range. However since the human and other gene sequences are now freely available, genome-wide searches should be made at the outset if sequence recognition targeting is being considered. Many types of sequence pattern are not randomly distributed, and may turn out to be under- or even over-represented.

In order for direct readout to occur, an incoming molecule has to access the base pairs via a DNA groove. The orientation of this incoming group is important since hydrogen bonding to and from bases, and hence direct readout, are predominantly directional interactions. The edges of the bases in Watson–Crick duplex DNA have hydrogen-bonding potential (Fig. 5.1) that is only partially satisfied by base pairing itself,[2] and therefore is available for recognition by ligands with hydrogen-bonding complementarity. This pattern of hydrogen-bond donation and acceptance has built-in sequence-dependency. Thus, in the major groove, a C•G Watson–Crick base pair has a pattern of (donor, acceptor,

acceptor) hydrogen bonding that is distinct from that for the reversed G•C base pair. The A•T and T•A base pairs have identical major groove donor/acceptor patterns, but the presence of the methyl group on thymine introduces an asymmetry in the groove that can, at least in principle, enable effective discrimination between these two base pair sequences. On the other hand, patterns of hydrogen bonding in the minor grooves of each pair of sequences are symmetric, so discrimination on this basis alone by molecules entering the minor groove is not straightforward.

There has been a perception that the minor groove is *always* narrow compared to the major groove in B-type DNA sequences. This has led to the view that it is unlikely that all three hydrogen bonds in a C•G or G•C base pair or the two in A•T or T•A base pairs, are simultaneously available for direct hydrogen-bonding sequence readout, for purely steric reasons. This accessibility is a consequence of groove widths, which are themselves sequence-dependent in both static and dynamic ways, as discussed in Chapter 3. It has been established from the large number of oligonucleotide crystal structures that A/T-rich minor grooves tend to be narrower than average; their widths depend both on the length and the nature of the AT sequence. However, these trends are by no means absolute. A more realistic view is that A/T regions are more flexible than G/C ones, and that particular circumstances such as the requirements of the spine of hydration, often result in narrowed minor grooves in A/T tracts of sequence. Other circumstances where the energetics of interaction override the stability of structured water, can readily alter this feature so as to markedly widen the groove. Consequently, it is possible to simultaneously access the minor-groove edges of both components of the base pairs in a run of A/T sequence with appropriate ligands that can stabilize a widened groove. Other sequence-dependent features of base pairs and base pair steps, which can be both localized (such as propeller and helical twist) and long-range (typified by the bending of A tracts), will affect the geometric relationships between hydrogen bond donors/acceptors on successive bases.

Indirect sequence readout by definition involves interaction with elements of DNA structure other than the base pair hydrogen bonds, and is consequently inherently less directional. The backbones themselves provide several mechanisms for such readout. Their sequence-dependent structural features have been outlined in Chapters 2 and 3. Advantage can be taken of differences in backbone and phosphate conformations such as B_I or B_{II} phosphate geometry, which can act as recognition signals. Perhaps more important are differences in distances between phosphate groups that necessarily result from backbone conformational differences, which can be recognized by hydrogen bonding/electrostatic interactions with basic amino acid side chains. The B-DNA minor groove in particular has extended regions of apolar character. Carbon and hydrogen atoms of sugar residues, notably C1′, H1′, C4′ H4′, C5′, and H5′ form a significant part of the van der Waals and solvent-accessible surface walls of this groove (Fig. 5.2). The apolar groove surface is interspersed at regular intervals by highly polar phosphate groups. Though fewer in number, there are some carbon/hydrogen atoms from base edges forming part of the groove floor surface, together with the polar

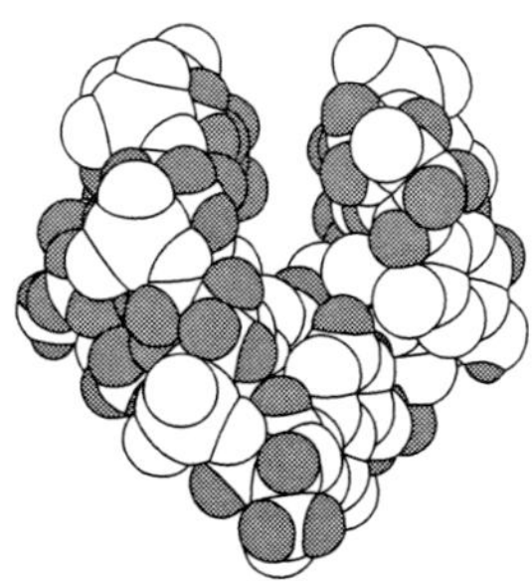

Fig. 5.2 A space-filling representation of a B-DNA duplex, looking down the minor groove, and with the hydrogen atoms in the groove shown shaded. Note that these form a significant part of the groove surface.

hydrogen bond donor/acceptor atoms on the edges. Groove depth is shallower in G/C regions on account of the exocyclic N2 substituent of guanine.

There are also significant differences in electronic character between A•T and G•C base pairs, with the latter being more electron-rich. Thus, electron-deficient planar molecules that can stack with individual base pairs (such as intercalating drug molecules or tryptophan side chains), will tend to bind preferentially to G•C base pairs. Such preferences are by their nature localized to individual base pair sites. There is a tendency for less stacking between bases in B-DNA duplexes for pyrimidine-3′,5′-purine dinucleoside steps than purine-3′,5′-pyrimidine ones (Fig. 5.3). The consequence of these differences is that the dinucleoside sequences CpG, TpG, CpA, and TpA (of pyrimidine-3′,5′-purine type) are more readily unstacked than their counterparts GpC, GpT, ApC, and ApT. The cost in energetic terms between them is only a few kcal/mole. However, this is sufficient for there to be a measurable preference for pyrimidine-3′,5′-purine base steps in the first group to be more readily unstacked and separated so that planar molecules of appropriate size can be inserted and bind intercalatively between them.

The methyl group of thymine can play an important role in major-groove sequence readout, by being able to make stable van der Waals non-bonded contacts with hydrophobic groups or side chains such as those of leucine, isoleucine, and alanine. Negative indirect readout can also take place when the presence of a particular group, substituent, or side chain would result in the occurrence of steric clashes with base substituents, and consequent destabilization of the resulting DNA–ligand complex. Examples include (i) the bulky methyl group of thymine driving major-groove recognition to a G/C region and (ii) the bulky exocyclic amino group at the 2-position of guanine, whose presence in

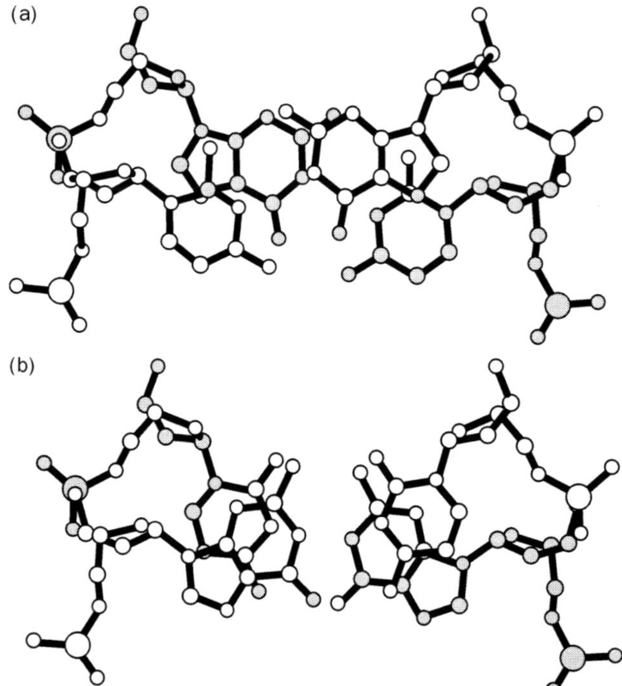

Fig. 5.3 Base stacking in a canonical B-DNA double helix at (a) pyrimidine-3′,5′-purine dinucleoside steps (b) purine-3′,5′-pyrimidine steps.

the minor groove acts as a deterrent to many minor-groove ligands. They then bind more readily in A/T regions.

The electronic characteristics of an individual base pair produce an electric field and potential in its vicinity. These electronic features are distinct for the various classical DNA structural types, as well as for changes in oligonucleotide sequences.[3] Of especial significance is the finding that for B-DNA an electrostatic potential minimum occurs in the major groove of oligo (dG)•(dC) sequences, and in the minor groove of oligo (dA)•(dT) sequences such as the 5′-AATT sequence in the Dickerson–Drew dodecamer. All these have greater negative electrostatic potentials than those around the phosphate groups, which explains why cationic ligands and basic protein side chains tend to cluster in the grooves rather than around the phosphates. These electrostatic potentials and fields are important in directing an incoming ligand to a region of sequence; small local variations in net charge and field around an individual base pair then could assume some importance, especially for electrophilically driven covalent bonding.

5.2 **DNA–water interactions**

Water plays a pivotal role in biological systems, in large part by use of the unique hydrogen-bonding ability of an individual water molecule. It can simultaneously act as acceptor and donor for up to a total of four hydrogen bonds, often forming extended networks, even in bulk water. The hydration of DNA is essential for the maintenance of its structural integrity. At least 12–15 first-shell water molecules are associated with each nucleotide in the B-form. Historically, it has been presumed that some of these are clustered around phosphate groups, together with counter-ions such as sodium, potassium or ammonium for charge neutralization. One would expect other water molecules to be associated with the bases, possibly taking advantage of the hydrogen-bonding facility (Fig. 5.1) offered by the edges of the bases.[1] Fibre diffraction analyses have achieved some success in locating the positions of some counter-ions, but only with the advent of single-crystal analyses of oligonucleotides has it been possible to define the positions of a significant proportion of the water molecules associated with a DNA sequence. NMR methods can provide valuable complementary information, by locating some of the least mobile water molecules, such as the minor-groove spine of hydration in the dodecamer d(CGCGAATTCGCG).[4,5] These methods can also provide data on residence times, which are typically in the low nanosecond range.[6] Crystallography, by contrast, provides a picture of hydration averaged over the time-scale of the X-ray experiment, which even for a high-flux synchrotron source, is many orders of magnitude greater than these times. Molecular dynamics simulations, which are increasingly being performed in this nanosecond range, can thus realistically model water motions around oligonucleotides. Recent simulations find excellent correlations between the majority of positions for statistical averaged waters from the simulations, and those found from crystallography.[7,8]

In general, the number of water molecules located by crystallography is highly dependent on the resolution of the structure and the thermal motion of individual water molecules (as well as on the quality of the raw X-ray diffraction data). It is necessary to achieve high resolution (≥ 1.5 Å) in crystallographic analyses in order to find more than a fraction of the water molecules beyond the first hydration shell. The widespread use of cryo-crystallographic methods to collect low-temperature diffraction data, especially at high-intensity synchrotron X-ray sources, is facilitating the location of an increasing fraction of these associated water molecules, since the freezing of crystals at liquid nitrogen temperatures enables all molecular motions to be decreased. In particular the motions will be reduced for quasi-bulk water molecules, which are involved in extensive water–water interactions. In favourable cases, complete second and even third-shell water molecules have been observed (see below). However, some water molecules are undoubtedly disordered, and only partially occupy discrete positions. These are harder to locate and differentiate from background noise in electron-density maps. The availability of database search and comparison methods has enabled systematic studies of water clustering to be made using the large number of oligonucleotide structures in the Nucleic Acid Database.[9,10]

The nature and extent of DNA hydration also provide an indication of accessibility to ligands other than water itself, which are borne out by calculations of the solvent-accessible surface characteristics of DNA. These have shown[11] that there are some differences between A and B helical types in terms of available surface area in the grooves, as well as between A•T and G•C base pairs. Sequence-dependent backbone conformational variability will also alter accessibility—when the minor groove in A/T regions is narrow its width is sufficient for the binding of only one water molecule per base pair in the first hydration shell. The first and second shells together produce a network of water molecules, the spine of hydration (see below), which is necessarily two-dimensional in form (Table 5.2).

The hydration patterns around phosphate groups in oligonucleotide crystal structures are dependent on whether the helix is an A, B, or Z type.[12,13] Analysis of water molecules around phosphate groups in all A-, B- and Z-DNA duplexes finds that each of the two charged oxygen atoms in a phosphate group can accommodate up to three waters of hydration.[10] The ester oxygen atoms, unsurprisingly, have little preference for hydration. In the A-form, successive phosphate groups along the phosphodiester backbone are close enough together for water molecules to be able to form bridges between them, as indeed is observed in many crystal structures. Such bridging patterns are not found in B-form oligonucleotides, with their more separated phosphate groups. These water molecules prefer to hydrogen-bond to the anionic phosphate oxygen atoms, whereas the ester oxygen atoms O3′ and O5′ are rarely hydrated. By contrast the deoxysugar ring oxygen atom frequently participates in water networks, especially in A-form structures. Several A-DNA octamer high-resolution structures have extended major groove networks of pentagonal-linked water molecules[14] whose coordination to bases appears to be a major factor in maintaining the integrity of the oligonucleotide structure.

Table 5.2 Oligonucleotide crystal structures showing particular hydration and ion features

Sequence	Feature	NDB ID code
d(CGCGAATTCGCG)	Spine of hydration	BDL002
d(CGCGAATTCGCG)	Spine of sodium ions	BDL084
d(CGCGAATTCGCG)	Potassium ions	BD0005, BD0041
d(CGCGAATTCGCG)	Calcium ions	BD0004

DNA-bound water molecules are frequently associated with metal ions, especially sodium, calcium, potassium, or magnesium. These ions, together with ammonium, are required for neutralization of the negatively charged phosphate groups, although only a small proportion of such ions are typically located in crystal structures. A provocative recent suggestion, discussed in detail below, has been that sodium ions can occupy part

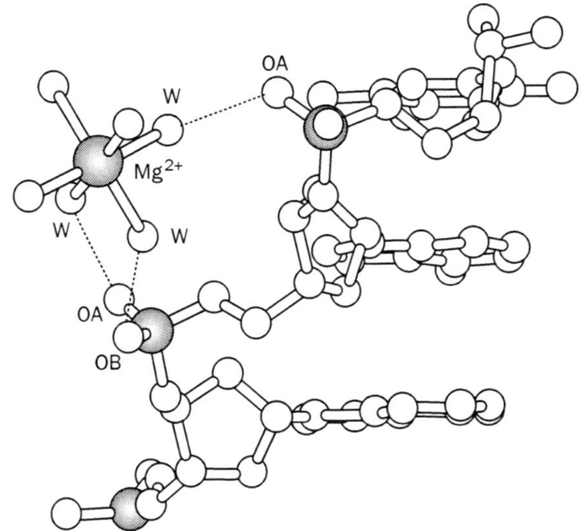

Fig. 5.4 Part of the crystal structure of a typical B-DNA oligo-nucleotide, showing a hydrated magnesium ion coordinated to and bridging phosphate groups.

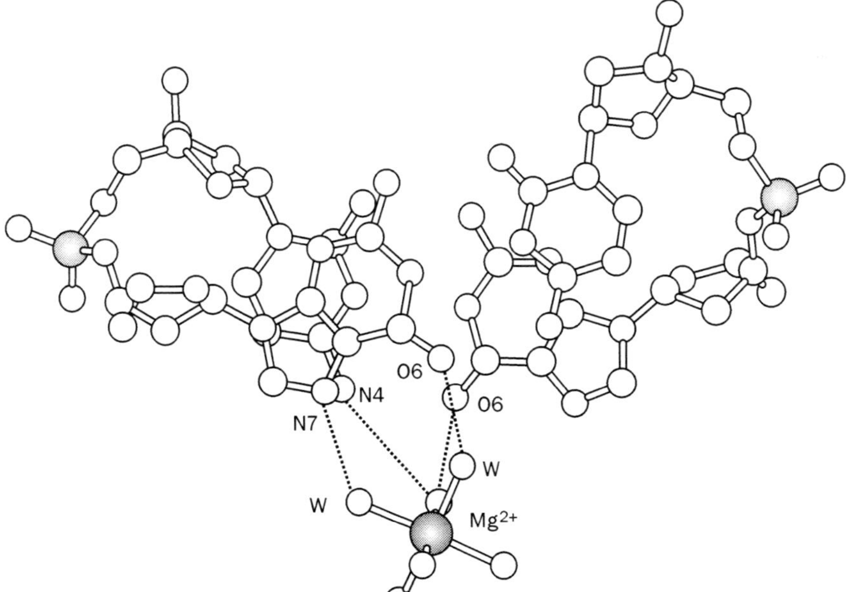

Fig. 5.5 Another view from the same crystal structure, where symmetry relationships in the crystal also place the hydrated magnesium ion in the major groove, where it can coordinate to atoms on base edges.

of the minor groove spine of hydration in B-DNA, possibly in fast exchange with water molecules. Magnesium ions are normally hexa-hydrated, and have been well located in a number of oligonucleotide crystal structures. Typically, their role is to bridge phosphate groups (Fig. 5.4). They can also lie in the major groove, contacting hydrogen bond acceptors at the base edges (Fig. 5.5), in both instances with hydrogen bonds from the waters of hydration associated with the magnesium ion.

5.2.1 Hydration in the grooves in detail

Systematic examination of the observed distribution of water molecules in known oligonucleotide crystal structures has shown that the dependence of individual base hydration on the helical type (A, B, or Z), is ultimately due to the size and electronic characteristics of the grooves in each type. For example, the B-DNA major groove is dominated by large, hydrophobic thymine methyl groups in A/T regions. The width of this major groove is such that water–DNA networks would have to be three-dimensional, even for the first shell of hydration. These networks would need to straddle between and contact the two backbones as well as contacting base edge donor/acceptors. It appears that such arrangements are looser than minor groove ones, and associated water molecules are more mobile. It is thus unsurprising that water molecules have only been observed to form extensive major-groove networks in high-resolution, low-temperature B-DNA oligonucleotide structures. Instead, the major-groove base hydration that has been observed in crystal structures is much more localized.[9] Water molecules tend to form discrete arrangements around particular bases, which sometimes extend to one or two bases in either direction. Purines frequently have water molecules hydrogen-bonded to them in the clustered manner shown in Fig. 5.6. Pyrimidines, by contrast, usually have a single associated water molecule, which can then hydrogen bond to another base, often via a second water molecule.

The hydration of the B-form minor groove has, and continues to be, the subject of continuing study and debate. The initial location[15] of the spine of hydration in the A/T region of the crystal structure of the sodium salt of the Dickerson–Drew duplex sequence d(CGCGAATTCGCG)$_2$ was at a resolution of 1.9 Å (see Chapter 3), which is only moderate by current standards. Nonetheless, the pronounced network of solvent molecules observed in this analysis is still accepted as essentially correct. This network involves a linear array of hydrogen-bonded molecules extending over more than six base

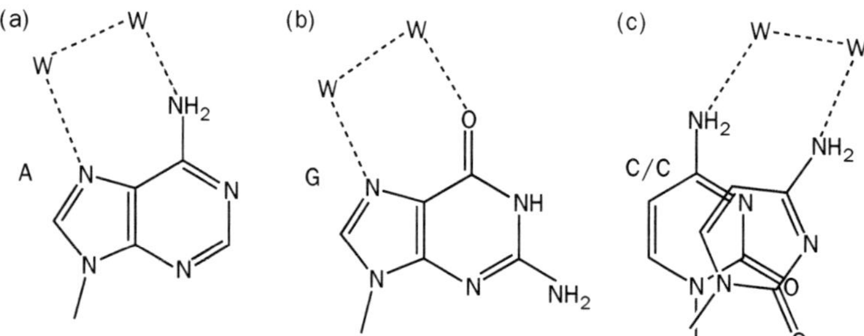

Fig. 5.6 Typical patterns of water clusters in a B-DNA major groove
(a) and (b) are around individual purine bases; (c) shows water molecules
bridging between two consecutive pyrimidines in the same strand.

pairs in the central region of the minor groove, with every other solvent molecule also hydrogen-bonded to a pair of base edge hydrogen-bond acceptors (Fig. 5.7). These, thus, bridge between adjacent base pairs.

The assignment of all these solvent atoms as water molecules has been re-examined in several more recent low-temperature studies of this structure, which also used crystallographic refinement methods based on more precise geometric parameters than were available to the original study. An analysis, extending to a resolution of 1.38 Å, has suggested that the primary shell of solvent (which directly contacts base pair edges), is partially occupied by sodium ions, and partly by water molecules.[16] These assignments are based on the behaviour of the solvent during the refinement, together with valence calculations. However, subsequent crystal-structure determinations, one at yet higher resolution (1.1 Å) using synchrotron data, have not found unequivocal evidence for sodium ion participation in the spine's structure.[17,18] The structure determinations of two newly discovered crystal forms of the dodecamer also concur with this view.[19–21] These analyses also show that minor-groove hydration extends well beyond the first- and second shells of the spine itself, with each group of three water molecules forming the nucleus of a hexagon of higher shell water molecules (Fig. 5.8). These eventually form hydrogen bonds to phosphate groups. The higher-shell solvent molecules in these hexagons are highly mobile, with large thermal factors. It is thus unsurprising that they have not been located in many studies.

Why is the unequivocal assignment of solvent molecules as waters or sodium ions so difficult? Since X-rays are diffracted by electrons in atoms, there are obvious problems of

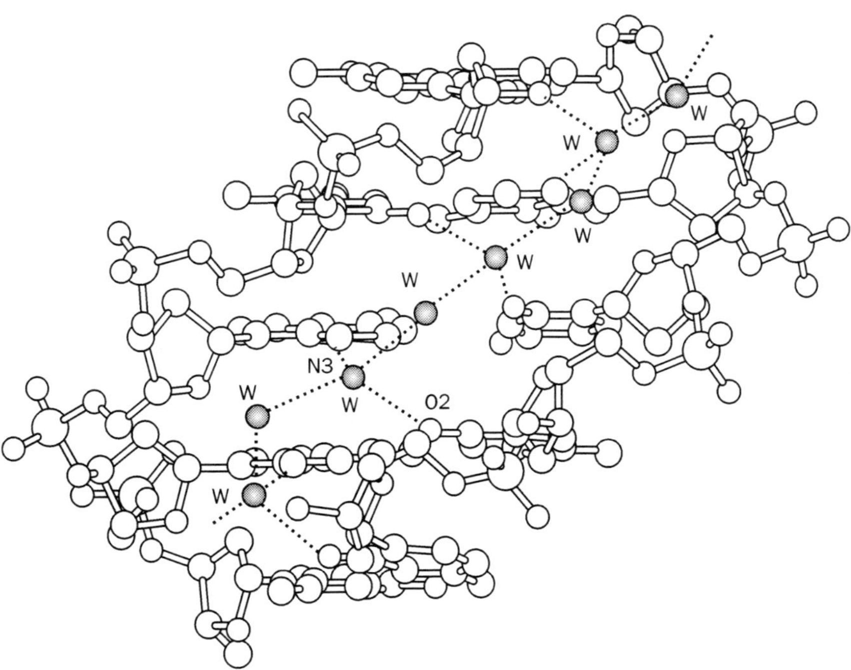

Fig. 5.7 A plot of the spine of water molecules in the Dickerson–Drew dodecanucleotide crystal structure.[15] The water molecules are shown as shaded spheres, and hydrogen bonds as dashed lines.

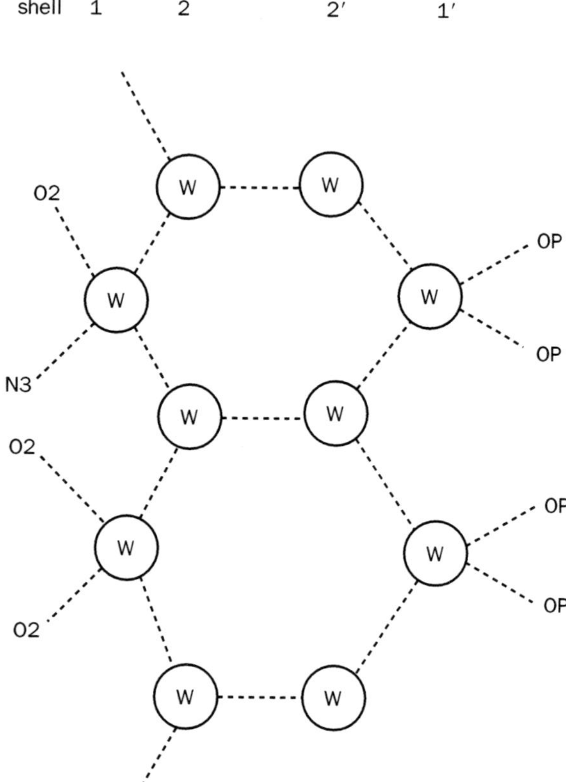

Fig. 5.8 A schematic representation of the arrangement of higher order shells of water molecules, as seen in high-resolution crystal structures of the Dickerson–Drew dodecanucleotide. The spine of hydration corresponds to shells 1 and 2.

differentiating one type of atom from another when the number of electrons in each is closely similar. An oxygen atom has eight electrons, and a sodium cation has ten. Even very high-resolution diffraction data has difficulty in distinguishing one from another. The problem is compounded by the mobility of solvent molecules, and by the probability that some have partial occupancies. On the other hand, discrimination between water molecules and potassium ions should be more straightforward with diffraction data extending to at least 2 Å. This is indeed the case.[22] X-ray crystallography cannot hope to differentiate water molecules from ammonium ions. NMR approaches using isotopically labelled ^{15}N ammonium ions have been used to show that they also partially occupy some minor groove sites, in fast exchange with water molecules.[23] The overall picture that emerges is one in which small solvent molecules continually aggregate in and dissociate from various buried sites (the grooves) and around exterior sites (the phosphate groups), on the average in geometrically defined ways. Water molecules play the major role, but ammonium and alkali metal ions can also come into play. Both small molecule and protein–nucleic acid complexes are also dependent on solvent for their stability. Water molecules can play an active role in the recognition processes in these complexes even though they still have short residence times.[24] Ions can play a significant role, in maintaining both local and large-scale conformational features such as A tract bending.[25]

5.3 General features of DNA–drug and small-molecule recognition

A large number of clinically important drugs and antibiotics are believed to exert their primary biological action by means of DNA non-covalent interaction and subsequent inhibition of template function. Cancer chemotherapy is in part based on such cytotoxic drugs, together with covalently binding agents. A well-studied example is the selective inhibition of DNA-directed RNA synthesis by the anti-tumour antibiotic actinomycin D. For many of these molecules, drug–DNA interaction *in vivo* involves concomitant interactions with the enzymes DNA topoisomerase I or II to form ternary complexes. This can lead to lethal DNA double-strand breaks, and eventually to selective tumour cell death, often via activation of the p53 apoptosis pathways. However, cytotoxic drugs generally have high toxicity to both normal and cancer cells, and their therapeutic index (the ratio between a therapeutic and a toxic dose) is usually poor. So, unsurprisingly, there has been a desire to rationally design new, improved ones, as well as to understand the molecular basis for the action of existing drugs. This has been the impetus for many of the large number of structural studies in this field. Almost without exception, these have been on binary drug–DNA complexes, and as yet there is little information at the atomic level on ternary complexes with protein partners.

DNA-interactive small molecules can be conveniently categorized into two major classes according to their mode of interaction. There are sub-divisions within each that frequently overlap. Electrostatic-dominated binding involves sequence-neutral interactions between a cationic group and negatively charged phosphates on DNA. Examples of such molecules include the natural polyamines such as spermine and spermidine. They are frequently used in nucleic acid crystallization trials in order to shield individual molecules from each other. It is rare for these polyamines to be located in electron-density maps. This suggests that, in common with alkali metal ions, their positions are mobile. The majority of non-covalent binding drugs carry formal positive charge, so electrostatic contributions are always a significant component of ligand binding.

5.4 Intercalative binding

There are a large number of antibiotics, antibacterial, and anti-tumour drugs that are characterized by the possession of an extended planar heteroaromatic ring system (chromophore). This essential structural feature is a chromophore of typically three fused six-membered rings in size (Fig. 5.9), which is approximately the same size as a base pair itself. The intercalation hypothesis, originally proposed by Lerman[26] suggests that the planar chromophore of the drug molecule becomes inserted in between adjacent

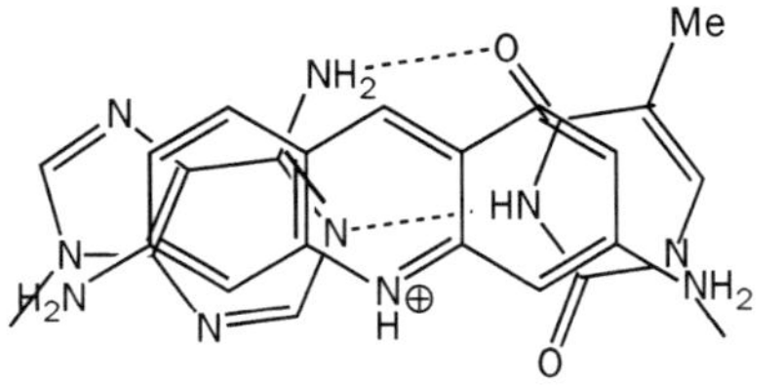

Fig. 5.9 The proflavine molecule superimposed onto a base pair, showing their similar sizes.

base pairs in an intercalative manner (Fig. 5.10). The drug chromophore is stabilized by polarization forces and van der Waals dispersion interactions between its planar group and the base pairs surrounding it. Optimal attractive interactions occur when the intercalating chromophore has polarized character, incorporating one or more hetero-atoms such as nitrogen. Electron-deficiency is also important, and a formal cationic charge is common (e.g. in the acridines). Intercalation results in an extension of the double helix by 3.4 Å per bound drug molecule, together with changes in helical twist (unwinding) for the base pairs at and adjacent to each binding site.[27] The base pairs either side of an intercalated drug thus increase in separation from 3.4 Å to 6.8 Å. The maximum number of drug molecules which can be bound to a DNA duplex is one for every three base pairs, that is, every other potential site becomes occupied at drug saturation. So the immediate base pair step either side of a binding site cannot bind a drug molecule. This is the neighbour-exclusion principle, which originates from the interpretation of solution binding data, but is based on the structural constraints forced on backbone geometry by intercalation.

Much of the interest in intercalating drugs has focused on their potential for anti-tumour activity. Some, such as the anthracycline antibiotic Adriamycin and its synthetic mimetic mitoxantrone, are of major clinical value in the treatment of human cancers. For the majority of these drugs and their derivatives, DNA binding abilities correlate approximately with biological activity. However, in spite of much activity in this field, the past 20 years has produced few new therapeutic agents with markedly superior activity. This situation may change when structural information becomes known for relevant drug–DNA–protein ternary complexes, especially those involving the topoisomerase family of proteins, which are often over-expressed in tumours.

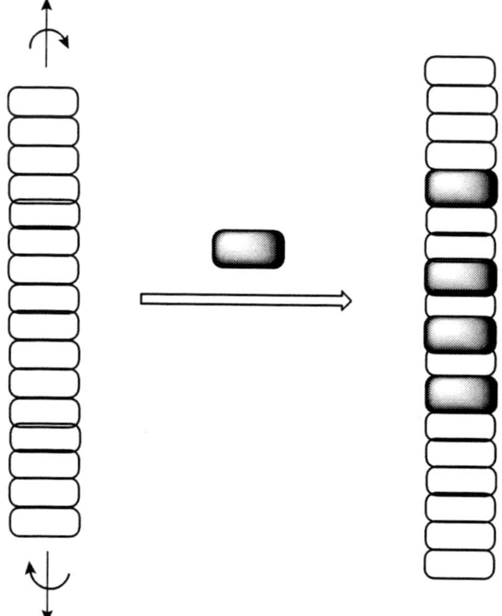

Fig. 5.10 A schematic representation of a planar drug molecule binding intercalatively into a DNA double helix, causing base pair extension and unwinding. Base pairs and drug molecules are shown, respectively, as open and shaded rectangles.

5.4.1 **Simple intercalators**

DNA-intercalator recognition itself is essentially sequence-neutral, although, as described in
Section 5.1, base pair stacking/unstacking requirements usually result in a small preference
for a pyrimidine-3′,5′-purine sequence step at the binding site. 'Simple intercalators' consist
solely or primarily of an intercalating chromophore, often carrying a positive charge on the
ring system—for example, proflavine and ethidium bromide (Fig. 5.11). The py-pu
sequence preference at the drug binding sites is generally obeyed both for these simple
intercalators and for the complex ones such as daunomycin and its analogues. For the
former, crystal structures of their complexes are restricted to DNA and RNA dinucleoside monophosphate 'mini' duplexes, provided the sequences are pyrimidine-3′,5′-
purine.[28] The complex intercalators, especially daunomycin and its analogues, have
been mostly studied structurally (Fig. 5.12) as hexanucleotide complexes, with again
drugs binding at the py-pu sites.[29] Analysis of conformational changes in intercalated
dinucleoside duplex structures and the observation of 'mixed' sugar puckers (C3′-*endo*-
3′,5′-C2′-*endo*), led initially to the suggestion that these are a necessary consequence of
the intercalation process, and would explain the neighbour-exclusion effect. This feature
was not observed in several crystal structures involving the acridine drug proflavine (such
as in ref. 28), and is no longer considered to be indicative of any fundamental structural
property. Final proof of this has come with analysis of the structure of a rhodium
ligand–DNA complex (see below).

The degree of helix unwinding produced by intercalation is dependent on the nature
of the bound drug. It can be measured experimentally[27] using covalently closed-
circular DNA, with reference to the standard value of 26° per bound drug molecule,

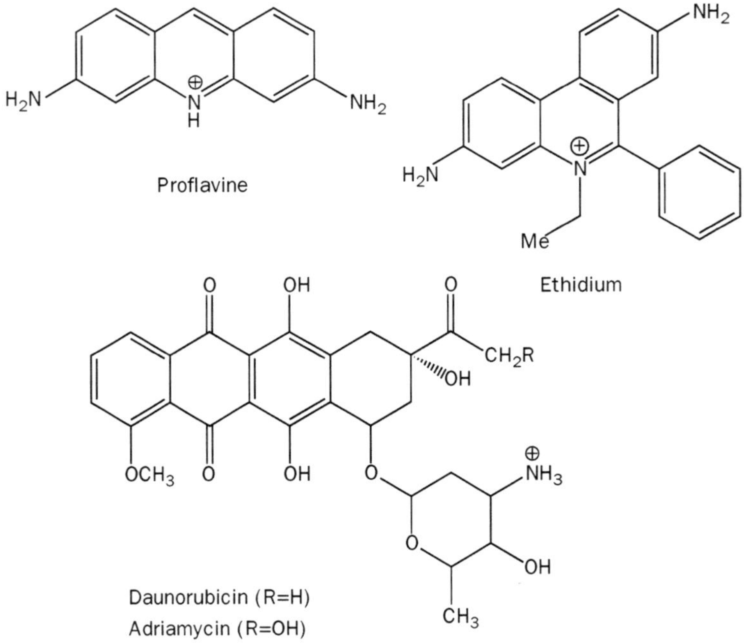

Fig. 5.11 The structures of some representative intercalating molecules.

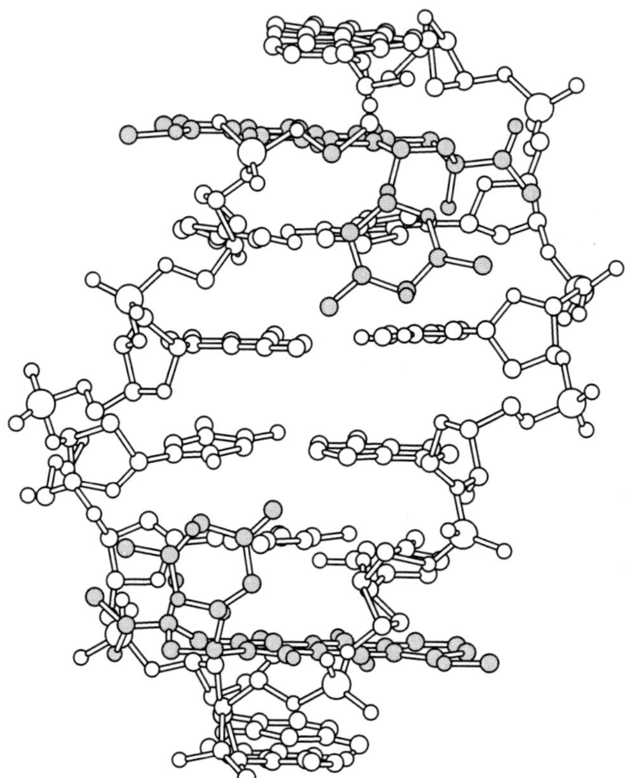

Fig. 5.12 The crystal structure of an intercalation complex involving the anticancer drug daunomycin bound at the terminal CpG steps of the hexanucleotide duplex d(CGTACG). The two drug molecules are shown with shaded atoms, and the structure is viewed into the minor groove, where the daunosamine substituent sugar rings reside.

for ethidium bromide. Crystallographic analyses of intercalation complexes have shown qualitative unwinding; the helical twist angle for the two base pairs immediately surrounding an intercalated drug, is not necessarily equivalent to this angle as measured in solution. The ribodinucleoside (CpG) duplex complex with proflavine[28] and the hexamer duplex d(CGTACG) with bound daunomycin[29] both have normal B-DNA helical twist angles of 36° for the base pairs flanking the drug. In the latter case, there are changes in helical twist at adjacent base pair steps. The unwinding angle is thus the sum of cumulative changes in helical twist over all affected base pairs. The process of base pair separation to produce an intercalation site also results in a number of other changes, especially in backbone conformation, and sometimes also in the details of base pair and base step morphology.

5.4.2 **Complex intercalators**

More complex intercalators than ethidium or proflavine have attached groups such as side chains, sugar rings or peptide units, for example, the anti-tumour drugs actinomycin D and daunomycin (Figs 5.12 and 5.13). These groups reside in a DNA groove, where they are stabilized by van der Waals interactions, and can hydrogen-bond to adjacent bases, providing sequence-specific direct readout. For example, the threonine residue in the cyclic pentapeptide part of the actinomycin molecule hydrogen-bonds to the N2

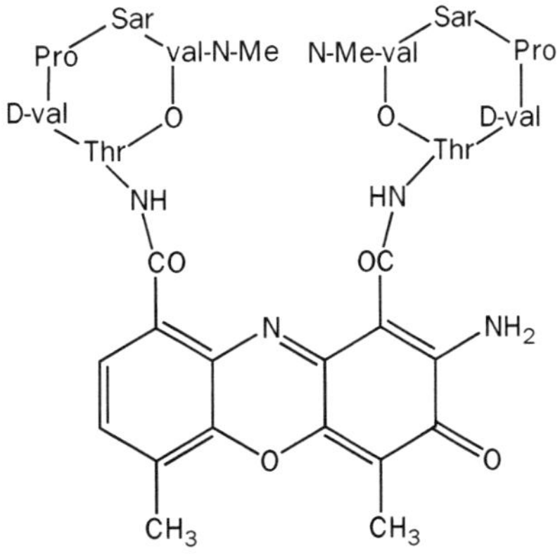

Fig. 5.13 The structure of actinomycin D.

Table 5.3 Crystal structures of drug–oligonucleotide intercalation complexes

Drug	Sequence	NDB ID code
Proflavine	d(CG)	DDB009
Ditercalinium	d(CGCG)	DDD030
Daunomycin	d(CGTACG)	DDF001
Daunomycin	d(CGATCG)	DDF018
Adriamycin	d(CGATCG)	DDF019
4′-Epi-adriamycin	d(TGATCA)	DDF035
Bis-daunomycin	d(CGATCG)	DDF072
Nogalamycin	d(CGTACG)	DDFA14
9-Amino-DACA	d(CGTACG)	DD0015
Triostin A	d(GCGTACGC)	DDH010
Actinomycin	d(GAAGCTTC)	DDH048
Rhodium complex	d(GUTGCAAC)	UD0005
Cu-TMPy	d(CGTACG)	DDF060
Ni-TMPy	d(CCTAGG)	DD0027
Psoralen	d(CCGCTAGCGG)	DD0030

and N3 atoms of a guanosine nucleoside, which is then structurally constrained to be on the 5′-side of the intercalating drug chromophore, giving a binding site requirement for the sequence GpX. This has been shown by a crystallographic analysis of actinomycin bound to the sequence d(GAAGCTTC) (Table 5.3). This structure shows that the phenoxazone chromophore of the drug is intercalated at the GpC site[30] (Fig. 5.14). The two pentapeptide groups are situated in the minor groove of the DNA helix, in agreement with an NMR study of an actinomycin complex with the sequence d(AAAGCTTT).[31]

NMR and crystallographic studies on oligonucleotide complexes of the anti-tumour antibiotic nogalamycin (Fig. 5.15) have revealed a novel binding mode for this drug, with

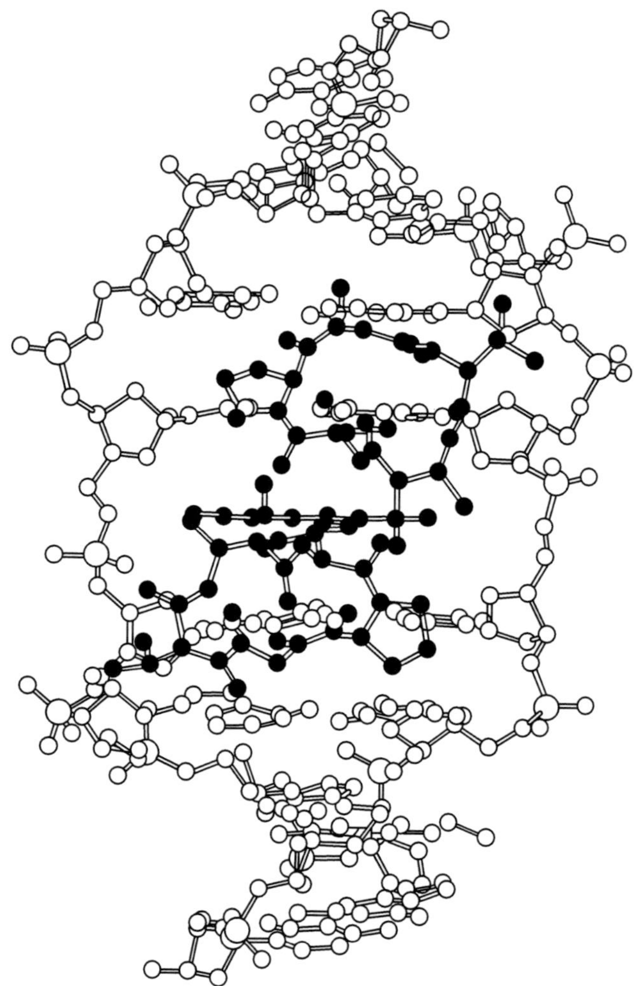

Fig. 5.14 Crystal structure[30] of a complex between actinomycin and the octanucleotide duplex formed by d(GAAGCTTC). Atoms of the drug molecule are shown shaded.

Fig. 5.15 The structure of the anti-tumour antibiotic nogalamycin.

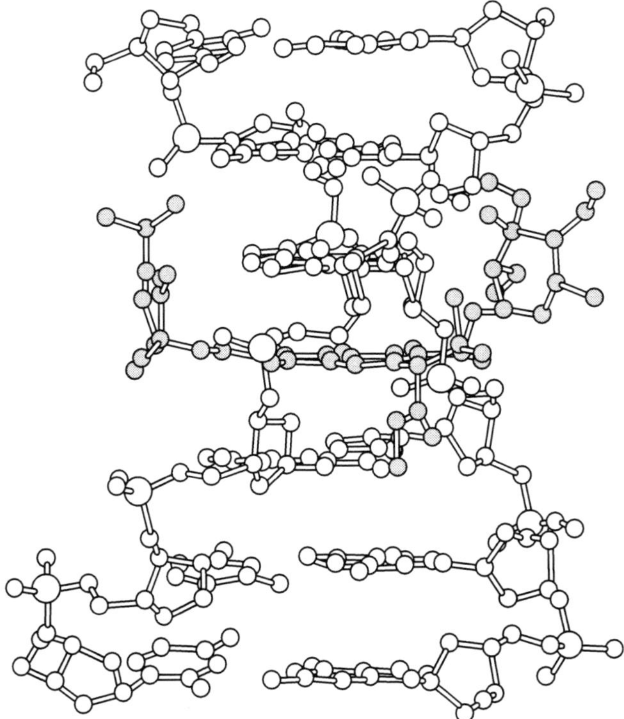

Fig. 5.16 The crystal structure[32] of the complex between nogalamycin and the duplex sequence d(CGTACG). The drug molecule is shown with shaded atoms.

the two attached groups residing one in each groove; the drug chromophore itself is intercalated as expected.[32,33] The aglycone group is held in the minor groove and the amino sugar in the major groove (Fig. 5.16), with drug–oligonucleotide hydrogen bonding to N2 and O6 atoms of guanines, thereby providing simultaneous major- and minor-groove direct sequence recognition, which agrees well with DNA footprinting data.[34] In contrast with the simpler intercalators, the mechanism of nogalamycin intercalation into duplex DNA is not straightforward. The two attached groups, one at each end of the molecule, are sufficiently bulky to ensure that the drug chromophore cannot become intercalated without one or other of these groups effectively passing through or opening up the intercalation site. This implies that drug–DNA association and dissociation will both be exceptionally slow processes as a direct result of the structural constraints forced on the complex, as has been found experimentally.

5.4.3 Major-groove intercalation

The majority of the 'complex' intercalating molecules have their attached groups bound in the minor groove, with the DNA duplex retaining B-like character. These drugs tend to be natural products and the pendant groups are then sugar residues or modified peptides, often with hydrophobic character which complement the hydrophobic areas on the groove walls. By contrast, a number of intercalating molecules have been synthesized which subsequent structural studies have shown to have their attached groups bound in the major groove. It is plausible to speculate that these groups would be especially effective in competing with regulatory proteins which have motifs inserted into the major groove of their target DNAs. Major-groove binding has been demonstrated for a family of synthetic anticancer drugs based on the acridine carboxamide skeleton. For example, the crystal structure[35,36] of a complex between the duplex d(CGTACG) and the 9-amino derivative of the parent drug in this series, termed 'DACA' (Fig. 5.17) shows that the cationic end of the dimethylaminoethyl side chain forms a hydrogen bond with O6 and N7 guanine base-edge atoms in the major groove (Fig. 5.18). Surprisingly, these drugs are also active against the enzyme DNA topoisomerase II, as are the minor-groove intercalators such as the anthracyclines, daunorubicin, and Adriamycin. This suggests either that the active site of the enzyme does not discriminate between major and minor groove drug substituents on a DNA complex, or that there may be distinct sub-sites for the two categories.

A number of metal-containing intercalating molecules have been synthesized as probes of DNA structure, and several are photo-active. The rhodium complex (Fig. 5.19) is typical. It was designed to recognize the major groove of the sequence 5'-TGCA, that is, intercalating unusually at the 5'-GC purine-3',5'-pyrimidine site. The remarkable crystal structure[37] of a complex with the recognition sequence embedded into an octanucleotide duplex has five independent complex molecules in the crystallographic asymmetric unit. They all show consistent structural features and conformations, with the planar group inserted from the major-groove direction, and the bulky non-planar part of the ligand fitting snugly against the walls and floor of the major groove (Fig. 5.20). This high-resolution

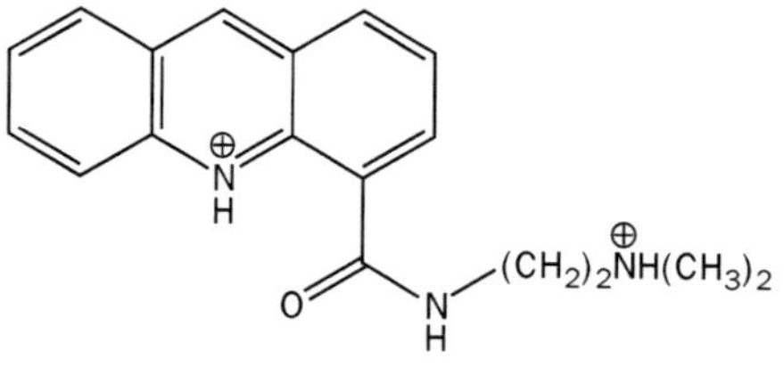

Fig. 5.17 The structure of the acridine anti-tumour drug DACA.

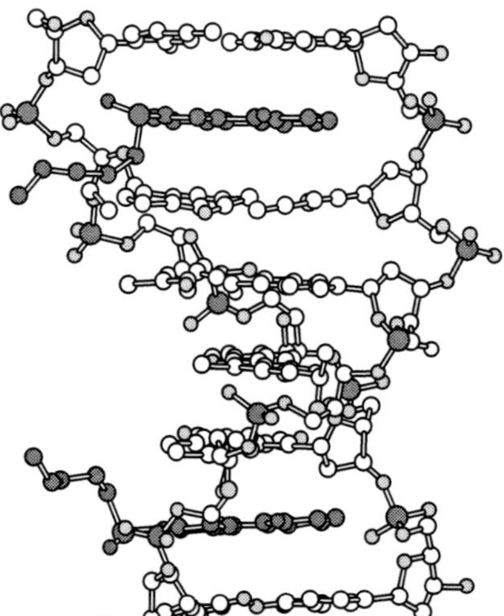

Fig. 5.18 The crystal structure[35] of a complex between 9-amino-DACA and the duplex sequence d(CGTACG), showing the side chains in the major groove of the DNA, positioned to hydrogen bond to base edges.

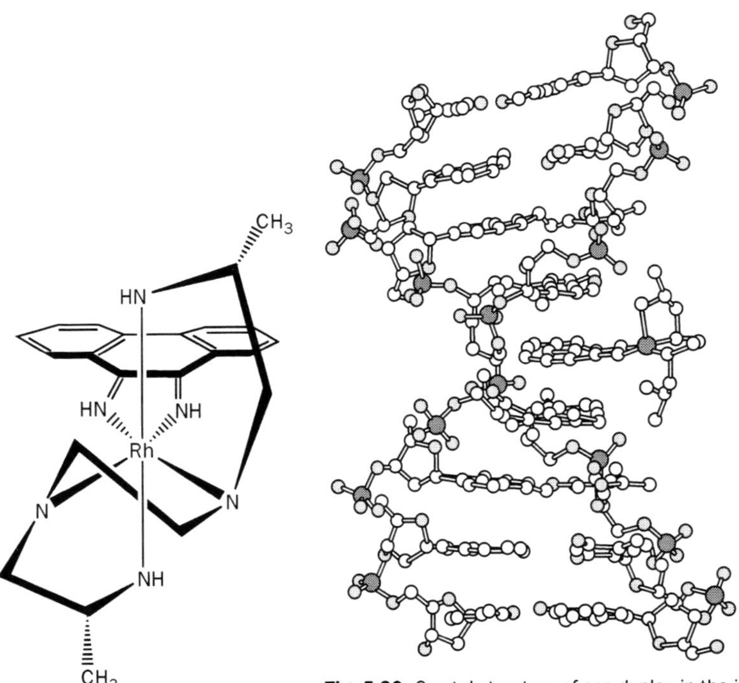

Fig. 5.19 Structure of an intercalating rhodium complex.[37]

Fig. 5.20 Crystal structure of one duplex in the intercalation complex[37] between the rhodium compound and an octanucleotide duplex, with the rhodium complex bound between the central base pairs.

structure is one of the few for an intercalation complex where the ligand is embedded at the centre of a significant length of DNA sequence, thus eliminating possible end-effects on conformation. It provides a view of the molecular features of intercalation that are relevant to the situation in bulk DNA. The long-standing controversy surrounding the nature of nucleoside sugar puckers at the intercalation site (see above) has been resolved, with the observation of consistent C2′-*endo* puckers at the ligand binding site.

A highly unusual and unexpected intercalation motif has been observed with the hexamer sequence d(CGTACG) in complexes with several diverse drugs and ligands—an acridine-4-carboxamide,[38,39] a bis-acridine and a synthetic daunomycin analogue, ametantrone.[40] In each case, conventional major groove intercalation into a duplex structure was expected. Instead, four duplexes in the crystal lattice come together to form an intercalation 'platform' with terminal cytosines displaced from one end of each duplex—this feature has been found in several native oligonucleotide crystal structure (see Section 3.3.2). Each displaced cytosine then effectively strand-invades another symmetry-related duplex so that a C•G base pair is formed. A pseudo-intercalation cavity is formed by two such adjacent duplexes, which has some resemblance to four-way junctions, and to ligand-binding sites in DNA quadruplexes. Whether this type of structure has biological relevance, or is solely an artifact of crystal lattice packing, remains to be determined. The crystal structure of a true four-way junction complex has recently been determined with a covalently bound derivative of the largely planar molecule psoralen.[41] This is further discussed in Section 5.5.

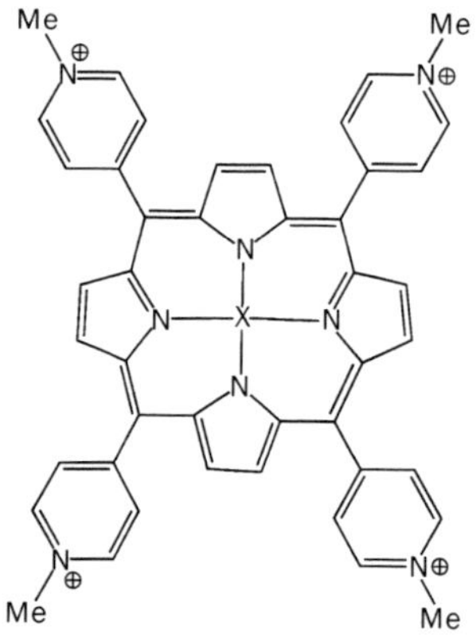

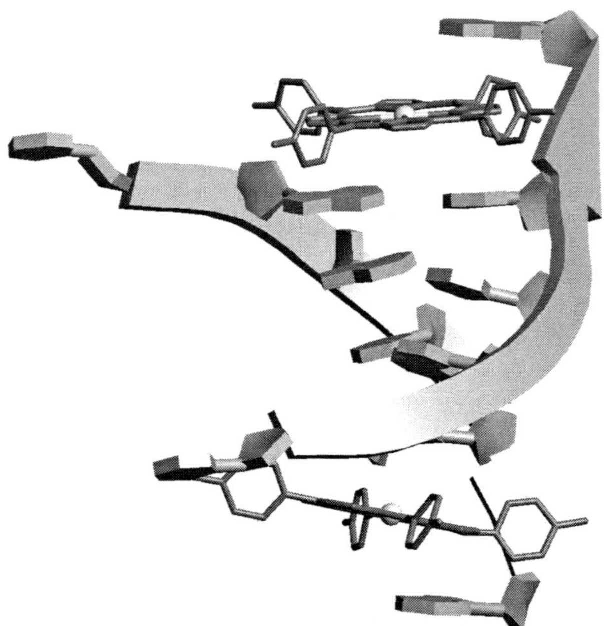

Fig. 5.21 Structure of tetra-N-methylpyridine-porphyrin. Atom X can be any one of a number of metal ions, such as nickel, copper or zinc. The native porphyrin molecule has hydrogen atoms attached to two of the four inner-facing nitrogen atoms.

The ultimate multi-groove binding molecules are the porphyrins such as tetra-N-methylpyridyl-porphyrin (TMPy) (Fig. 5.21). Molecular modelling studies with this ligand have predicted that the four substituents could fit into a B-DNA intercalating site with two substituents in each groove. Its sequence preferences have been interpreted in terms of minor-groove binding at A/T regions and intercalative in G/C ones. Intercalation has been inferred by NMR methods in an oligonucleotide complex to take place at the sequence CpG, in accordance with the pyrimidine-3′,5′-purine sequence preference rule.[42] This would necessarily place two of the non-planar N-methylpyridine substituents in each groove, major and minor. However, the crystal structure of a complex with the hexanucleotide d(CGATCG) duplex and a copper complex of this porphyrin does not show full intercalation.[43] Instead, the porphyrin is best described as being semi-intercalated at the two terminal CpG sites, with the end cytosine in each case being swung out of the helix (Fig. 5.22). Thus, the overall amount of porphyrin-base stacking is less than expected for full intercalation. The nickel complex of the same porphyrin forms a quite distinct complex[44] with the sequence d(CCTAGG). The hexanucleotide forms a B-DNA duplex. However, the porphyrin molecule, which is significantly buckled out of planarity, is not intercalated, but stacked at the ends of a pseudo-dodecanucleotide formed by two

Fig. 5.22 The crystal structure[43] of the copper complex of TMPy bound to the sequence d(CGTACG). The copper atom is coloured black.

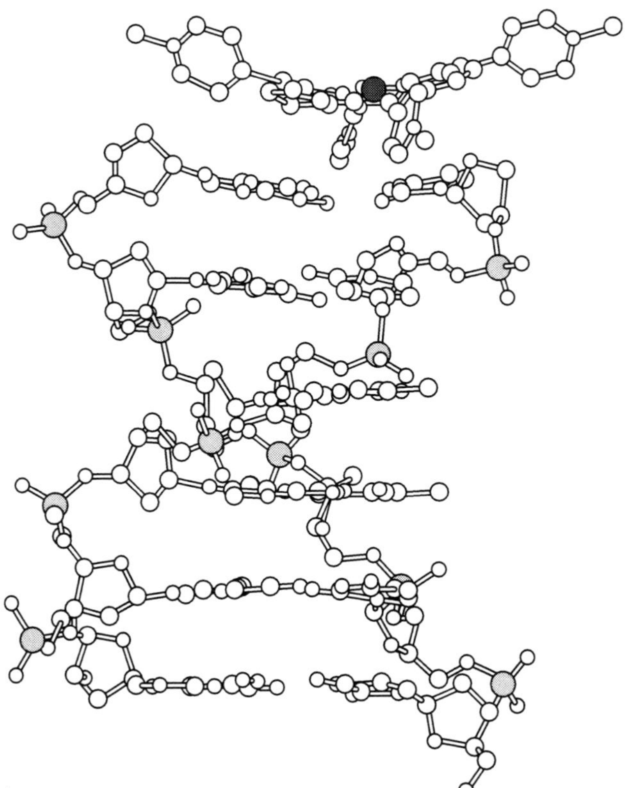

Fig. 5.23 The crystal structure[44] of the nickel complex of TMPy bound to the sequence d(CCTAGG). The nickel atom is coloured black.

end-to-end hexanucleotide duplexes (Fig. 5.23). The crystal packing of this structure positions one such dodecanucleotide duplex with its porphyrin end-caps, against the minor groove of another, with the effect of placing one porphyrin end-cap in the widened minor groove of the duplex, thereby providing a visualization of the groove-binding mode of this ligand.

5.4.4 **Bis intercalators**

Yet more complex are the drug molecules typified by the echinomycin and triostin family of anti-tumour antibiotics (Fig. 5.24), which have two intercalating chromophores linked together by cyclic oligopeptides. These molecules intercalate such that the two chromophores, whose planes are separated by a distance of 10.2 Å (i.e. three base pairs), can simultaneously bind at sites separated by two intervening base pairs (Fig. 5.25). This mode, termed bis-intercalation, has been validated by crystallographic[45,46] and NMR[47,48] studies on echinomycin and triostin A bound to oligonucleotide sequences. Bis-intercalator can also be characterized by unwinding experiments on closed-circular DNA, since they produce approximately twice the unwinding per bound drug molecule compared to a mono-intercalator. Echinomycin itself has a sequence requirement for the dinucleoside 5′-CpG, with two C•G base pairs being flanked by the two quinoxaline chromophores. The peptide ring system sits in the DNA minor groove, with specific hydrogen bonding to the 2-amino group of the guanine. An unexpected finding in the

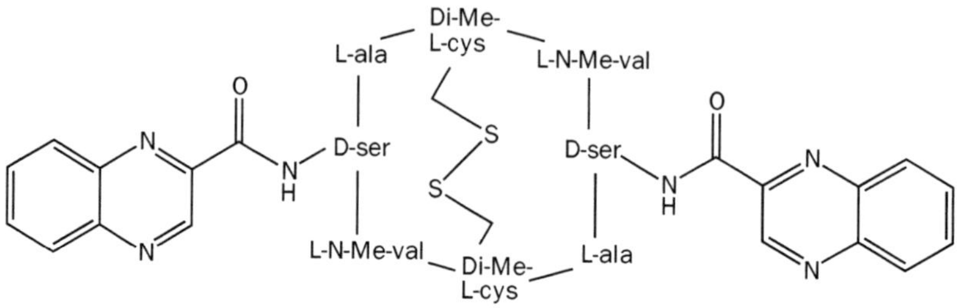

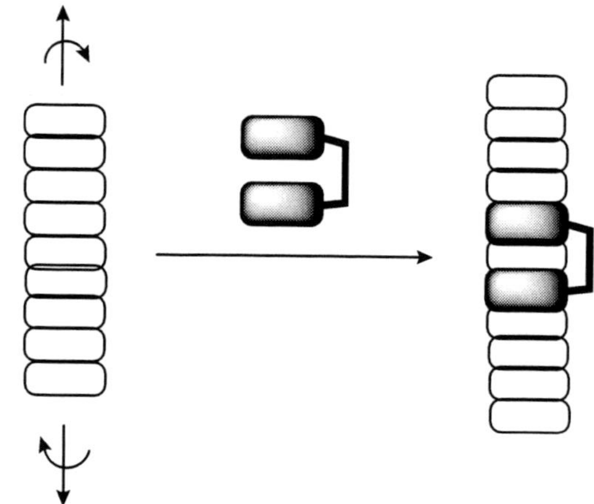

Fig. 5.24 The structure of triostin A.

Fig. 5.25 A schematic view of bis-intercalation.

crystal structures (Fig. 5.26) of the drug complexes has been the occurrence of Hoogsteen hydrogen bonding for the base pairs 'externally' flanking the bound drug—the two base pairs between the chromophores are always in a standard Watson–Crick arrangement. NMR studies indicate that this major structural change from standard B-DNA can occur with these drugs and certain short oligonucleotide sequences in solution. However, chemical probe experiments[49] with much longer (160 base pair) sequences that are more truly representative of biological DNA, strongly suggest that Hoogsteen base pairing does not occur in these sequences. It appears then that Hoogsteen pairing does not always need to be formed as part of the structural changes in a DNA duplex consequent to bis-intercalative recognition, and that these changes may be dependent on sequence length.

A novel series of synthetic bis-intercalating analogues of daunomycin have been developed by a structure-based drug design programme,[50] based on two rationales:

(1) such molecules would be superior DNA binders compared to daunomycin and its analogues since their two chromophores would produce greater stacking interactions;
(2) it is frequently found in the anthracycline series of compounds that intercalative DNA-binding affinity correlates with cellular cytotoxicity that in turn can indicate trends in anti-tumour activity. Thus, it was anticipated that bis-daunomycins would show enhanced biological activity compared to their mono-analogues.

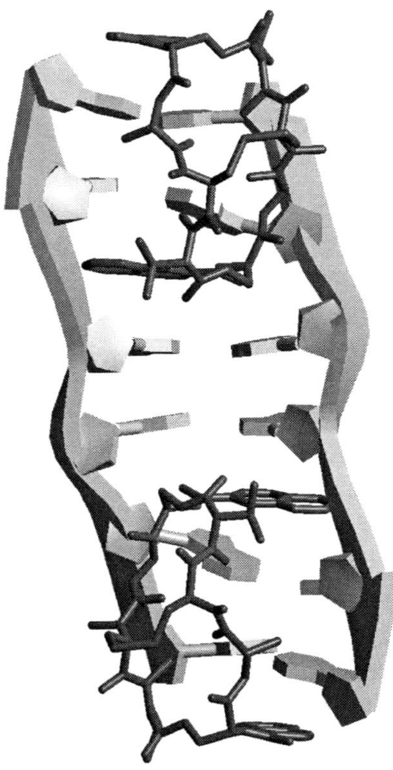

Fig. 5.26 The crystal structure[45] of a complex between an octa-nucleotide duplex and two triostin molecules, each bound in a bis-intercalative manner.

The standard hexanucleotide-daunomycin crystal structures,[29] as described above, were taken as a starting-point, with two drug molecules bound in the hexanucleotide duplex. The two minor-groove pendant sugar groups were replaced by a p-xylene linker group through the exocyclic amino groups on each sugar ring, thus generating a single bis-intercalating molecule. The resulting molecule 'WP631' (Fig. 5.27), has been shown to exhibit all of the enhanced DNA binding properties predicted for a bis-intercalator. Both NMR and crystallographic studies (Fig. 5.28) have verified this binding mode.[51] The DNA affinity of WP631, at 10^{11} mol^{-1}, approaches that of many regulatory proteins. However this is not reflected in a quantum leap in biological response since its anti-tumour activity is not greatly superior to that of the native daunomycin. So, in spite of the elegance of the drug design process in this instance, there appears to be an inherent limitation to the efficacy of intercalating molecules as anti-tumour agents. This is really unsurprising since molecules such as WP631 will have high toxicity to normal cells, and are not capable of being targeted to any oncogenes.

Bis-intercalation from the major-groove direction is rare. It has been observed[52] in the ditercalinium molecule, comprising two pyridocarbazole units, linked by a flexible aliphatic chain, which has been co-crystallized with the short duplex sequence d(CGCG). Each planar group of the ligand is intercalated at the CpG step, and the linker is positioned in the major groove. There is a marked bend in the DNA, towards the minor groove, which may be an innate feature induced by this type of bis-intercalating molecule.

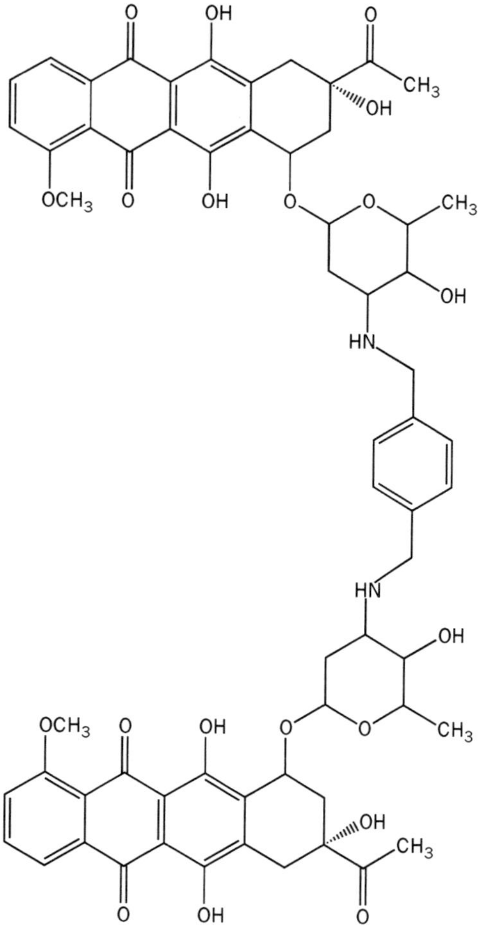

Fig. 5.27 The structure of the bis-daunomycin molecule, WP631.

5.5 **Intercalative-type binding to higher order DNAs**

Binding of small molecules to triplex DNA is a well-established approach to stabilize tracts of triple-stranded sequences.[53–57] The effective molecules in use have the common features of planar aromatic chromophores containing several fused electron-deficient rings, and it is presumed that they intercalate into the triplex at some point. A number of them show preferential binding to triplexes over duplex DNA. Stabilizing ligands studied include benzo[e]pyridoindole, the alkaloid coralyne and di-substituted amidoanthraquinones. There is little direct structural data on these complexes, although several molecular modelling studies have successfully rationalized binding behaviour in terms of intercalative-type models (Fig. 5.29). Selectivity for triple-stranded DNA arises from a combination of base-chromophore stacking and substituent binding in triplex grooves.

The first report of drug binding to a guanine-quadruplex type of DNA sequence involved the simple intercalator ethidium,[58] although it was not possible to define the mode of binding. This report lay dormant in the literature until recently, since when

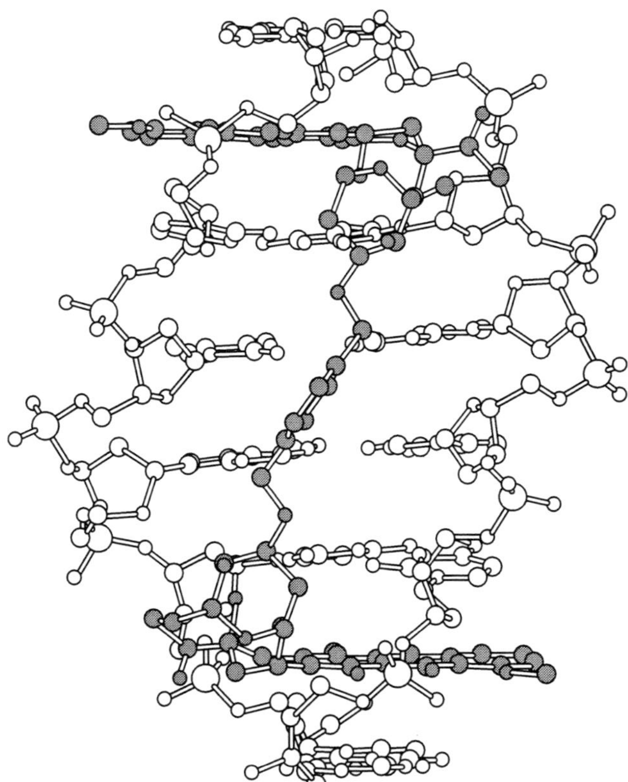

Fig. 5.28 The crystal structure[51] of the bis-daunomycin molecule WP631 bound to the duplex sequence d(CGTACG), viewed towards the minor groove and showing the linker group covering four base pairs.

there has been considerable interest in such four-stranded structures formed from telomeric DNA sequences, and their complexes with a range of intercalative type of molecules.[59,60] This activity has arisen as a consequence of the findings that telomeric DNA is progressively shortened in normal cells, but is stabilized in length in most tumour cells by a specialized reverse transcriptase enzyme complex, telomerase.[61,62] This requires the telomere strand, acting as primer, to be single-stranded in order to hybridize with the RNA template component of the telomerase complex. Folding the primer strand into a quadruplex arrangement effectively inhibits telomerase.[63] This folding process can be promoted by a range of intercalator-type molecules. Various substituted anthraquinones, acridines and the tetra-N-methyl-pyridyl porphyrin all show correlations between quadruplex binding and telomerase inhibition.[64–66] The majority of these molecules are also effective duplex binding agents, and thus show little selectivity for G-quadruplexes.

That it is possible to generate selective G-quadruplex-interactive molecules has been shown by several studies. Whereas ethidium itself binds only weakly to quadruplexes, derivatives with appropriate side chain functionality have been shown to bind with high affinity, and to be potent telomerase inhibitors, as have a number of dibenzophenanthroline derivatives[67,68] and tri-substituted acridine compounds. These latter molecules have been successfully designed to exploit the groove differences between quadruplexes and duplexes, and are also more effective telomerase inhibitors.[69] This field is still in its infancy, with the diversity of possible quadruplex-type arrangements for G-rich (and C-rich) sequences providing a complex set of possible targets to be explored for drug action. As yet there

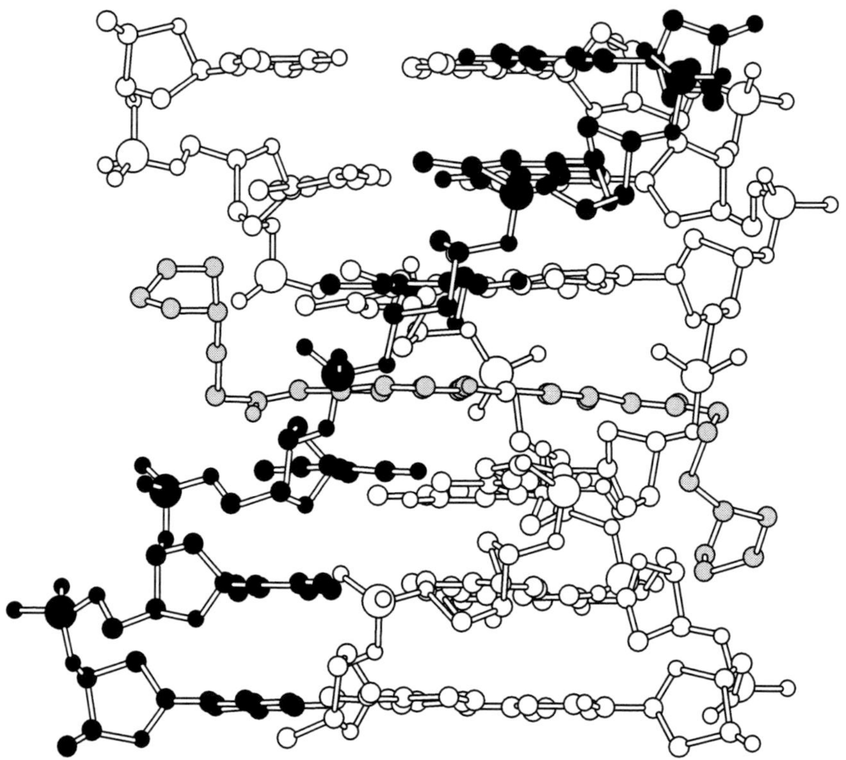

Fig. 5.29 A molecular model of a disubstituted anthraquinone (shown with shaded atoms) intercalated into a B-type parallel triplex structure. The third strand of the triplex B highlighted.

is little detailed structural data on ligand complexes, although most NMR, chemical cleavage and molecular modelling studies conclude that intercalative-type binding does not occur between stacked G-quartets, but rather ligands are externally stacked at the ends of quadruplexes.[70,71] In the case of the human four-repeat intramolecular quadruplex, molecular modelling has suggested that the major (intercalative-type) binding site is between the diagonal T_2A loop and the G-quartet at the 5′-AG step. G-rich sequences are mostly located in telomeres, although some promoters (e.g. that of the *c-myc* oncogene), also contain putative quadruplex-forming DNA, albeit in duplex form. The recent finding[72] of an association between the C-rich strand and the quadruplex-binding ligand tetra-N-methyl-pyridyl porphyrin, shows that the i-motif may also be of significance as a drug target. A model for this complex has been suggested from NMR data, with the ligand bound externally to an i-motif tetraplex. This is analogous to the NMR structure for this porphyrin complexed with a G-quadruplex,[65] and both bear some resemblance to the crystal structure of the nickel porphyrin stacked on the ends of a d(CCTAGG) duplex.[44]

The crystal structures of two DNA four-way Holliday junctions complexed with a derivative of the covalently binding intercalating drug psoralen,[41] are remarkable demonstrations of the power of particular DNA sequences to form non-duplex structures. One structure, with the sequence d(CCGGTACCGG), is not unexpected, given that this sequence forms a junction structure in the absence of ligand—the complex is nearly identical to the native structure. By contrast, the related sequence d(CCGC-TAGCGG) dimerizes to a duplex in the absence of psoralen, but is induced by the drug to form a related (though non-identical) junction structure. It has been suggested that this is a model to explain the promotion of recombination events by psoralen, leading to repair of the resulting lesions in DNA. These structures raise the intriguing question of whether other, non-covalently-binding drugs, can similarly promote and stabilize four-way junctions, since the central junction regions in them have considerable free space.

5.6 **Groove binding molecules**

A large and chemically diverse family of compounds can be classified as DNA groove-binders. Many show biological activity, and several of them find medicinal use as anti-parasitic or anti-viral agents. They generally show a preference for binding to A/T regions of DNA.[73,74] By contrast with intercalating drugs, they do not significantly perturb DNA structure. They bind exclusively in the minor groove of B-DNA duplexes. They can function as simple blockers of transcription, or as inhibitors of DNA topoiso-merase enzymes. This alone is probably insufficient to explain why some minor-groove binders work as drugs. It has been suggested that for those drugs with high therapeutic indices in diseases where an external organism is the causative agent (e.g. in microbial infections), preferential drug binding would occur in extended regions of A/T sequence in the organism. Such regions have been found in the mitochondrial DNA of these organisms and the drugs may, therefore, be selectively inhibiting their electron transport functions. Groove-binding ligands are also attracting increasing interest as starting-points for the design of sequence-specific molecules capable of the recognition of unique sequences within a genome (see Section 5.1). These would have general applicability to a wide range of human diseases.

5.6.1 **Simple groove binding molecules**

Molecules which have been shown to bind preferential to A/T-rich duplex DNA, but not to duplex RNA or A-form DNA include synthetic DNA stains typified by Hoechst 33258 and DAPI, and anti-trypanocidal agents such as berenil and the drug pentamidine (Fig. 5.30). Biophysical and footprinting studies have demonstrated that these molecules bind with approximately the same affinity to DNA as intercalators (with typical binding affinities of 10^6 mol^{-1}), but do not perturb DNA structure.

The common structural characteristics shared by most minor-groove binding mol-ecules are:

- positive charge(s)
- linked rather than fused aromatic and/or heteroaromatic rings
- an approximately crescent shape in three dimensions.

A number of X-ray crystallographic and NMR studies have shown that these molecules bind into the minor groove of B-type DNA duplexes, and that these common features play an important role in the interactions (refs 75–78).

Almost all these structures involve complexes with duplexes formed by dodeca-nucleotides, most often with the Dickerson–Drew sequence d(CGCGAATTCGCG) and closely related ones such as d(CGCAAATTTGCG). Typically, the ligand is bound in

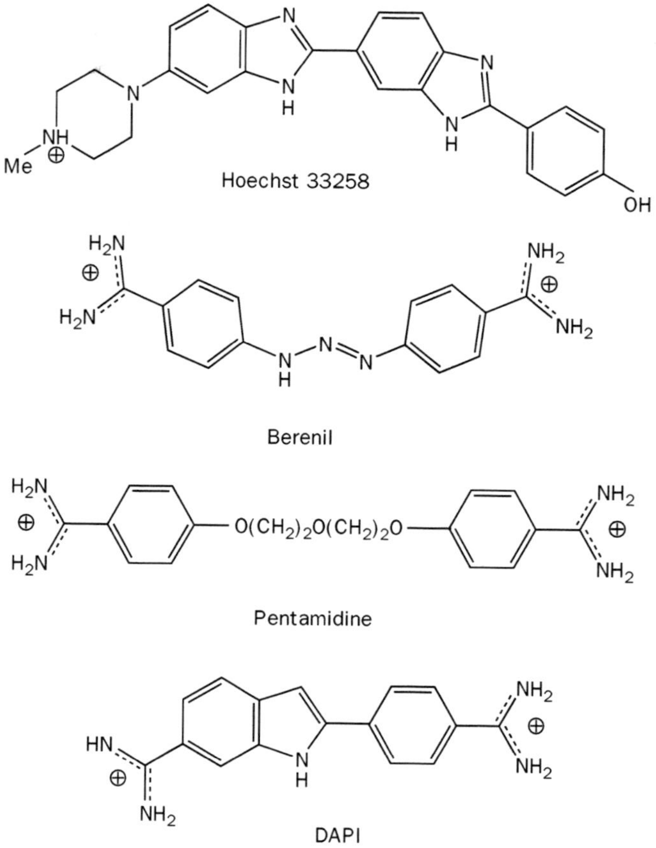

Hoechst 33258

Berenil

Pentamidine

DAPI

Fig. 5.30 Some typical groove-binding molecules.

the A/T region, with the aromatic groups of the ligand lying between and parallel to the two sugar-phosphate backbones (Fig. 5.31). There is little change to the structure of the DNA upon groove binding, at least for these ligands. The groove tends to be exceptionally narrow in this region, and the backbones and groove floor are in close van der Waals contact with the ligand. Much of this is hydrophobic in nature, and predominantly involve non-polar atoms C1′/H1′, C4′/H4′ and C5′/H5′ of the backbone, as seen in Fig. 5.2. The chemical shifts induced in these protons on binding can be taken as diagnostic for groove binding. Hydrogen bonding is frequently observed, notably between a donor atom on a ligand hetero ring or one in a charged terminal amidinium group (Fig. 5.32). Such hydrogen bonding is to a thymine O2 or an adenine N3 atom. This is termed direct sequence readout, and is highly directional. Conversely, sequence readout can also be indirect, with the ligand being able to recognize particular backbone or other sequence-dependent structural features. For A/T sequences indirect readout is a consequence of (i) the narrow cross-section of groove-binding molecules complementing groove width and (ii) the negative electrostatic potential in the AT minor groove complementing the positive charge(s) on these molecules. Groove interaction also involves the concave curvature of the inner surface of the bound molecule complementing that of the convex surface of the floor of the DNA minor groove itself. Thus, a G/C sequence with an exocyclic NH$_2$ amino group of guanine protruding into the groove will hinder the effective binding of a molecule with a smooth concave inner surface. This surface matching has been termed 'isohelicity'[79] since both the drug inner surface and the floor of the minor groove have twists in their curvature as a result of the helical nature of the DNA double helix. Isohelicity (Fig. 5.33) has been found to be a useful concept in the design of novel groove-binding agents.

There has been controversy concerning the relative importance of the various factors contributing to sequence-selective binding. A number of ligands have been shown to have minimal or even no hydrogen bonding to base edges, and electrostatic factors are of lesser importance in those instances where the ligands are uncharged. van der Waals and hydrophobic interactions with groove walls are probably the dominant factors in overall stability when ligand is bound in the groove, and is driven in part by groove width structure and flexibility. Directed hydrogen bonding, though a smaller contributor to overall stabilization, is responsible for directing binding to particular sites and sequences within an overall sequence type. Ligands that have a general A/T selectivity and do not discriminate between different types of A/T sequence, are not always involved in significant hydrogen bonding to bases. Hydrogen bonding is sometimes a crucial factor and can be exploited in the design of molecules with altered sequence recognition capabilities—see below. Groove

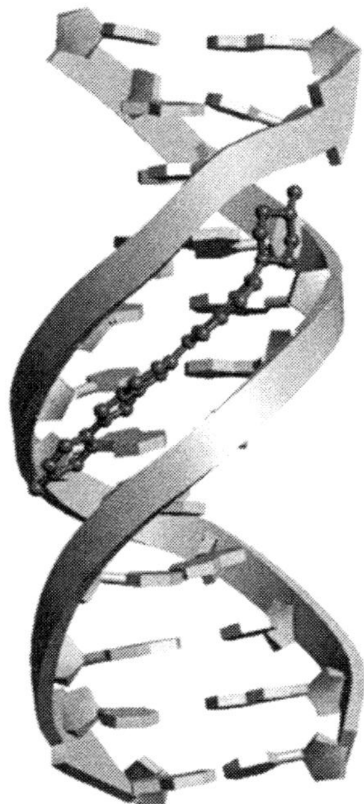

Fig. 5.31 A view of the crystal structure of the complex between Hoechst 33258 and the duplex formed by d(CGCAAATTTGCG), with the atoms of the drug molecule coloured black.

binding at A/T sequences displaces the highly structured arrangement of water molecules, the 'spine of hydration'. The energetic driving force for this displacement is the dominance of the non-polar interactions between ligand and groove walls. Variations in groove width are probably responsible for the greater affinity of molecules such as berenil, netropsin, Hoechst 33258, for sequences containing 5′-AATT compared to 5′-TTAA or 5′-TATA.[80,81]

A well-studied sub-class of groove binders comprise molecules with two charged amidinium groups, one at each end. Berenil is typical, with a triazine group linking two phenyl amidinium moieties. Crystallographic and NMR analyses have shown that the molecule covers 3–4 A•T base pairs.[75,76] The former also find that each amidinium group has an N–H bond facing into the groove, which hydrogen-bonds with thymine O2 or adenine N3 base edge atoms. In the complex with the Dickerson–Drew sequence, a water molecule mediates this interaction at one end of the binding site. A number of other bis-amidinium drugs, notably those with a flexible linker such as the drug pentamidine, bind in a similar manner. Pentamidine is of considerable clinical importance, since it has activity against the *Pneumocystis carinii* pathogen, which is responsible for the strain of opportunistic and life-threatening pneumonia that affects the majority of AIDS patients. The drug probably works by inhibiting the topoisomerases of the pathogen, via binding to the A/T-rich genome at points of topoisomerase selectivity. Although pentamidine is one of the drugs of choice, its effectiveness is limited and it produces a number of toxic side-effects. Accordingly, there have been a number of studies directed to finding more

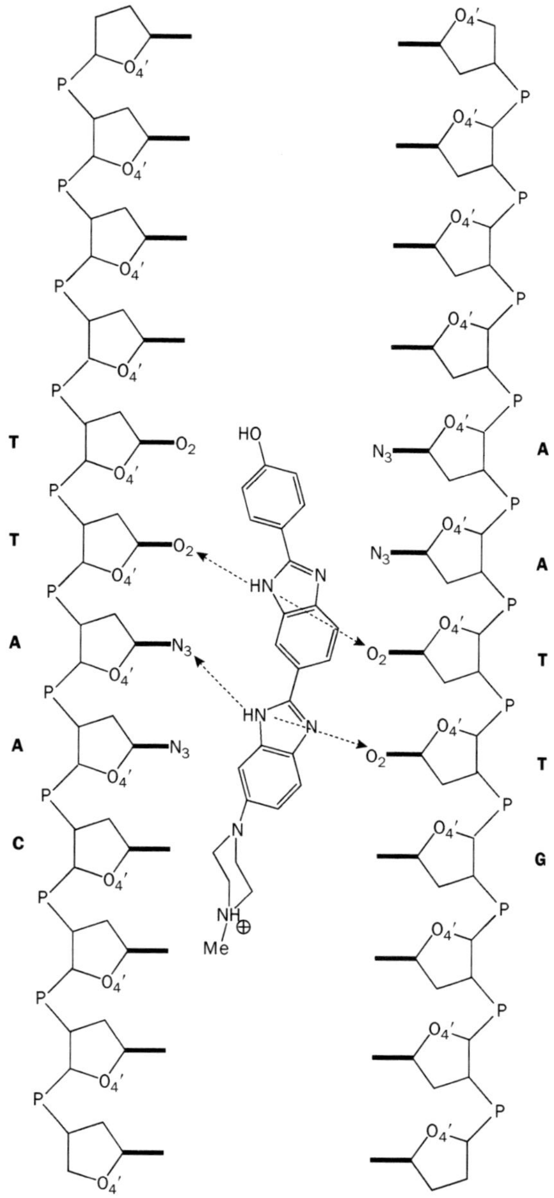

Fig. 5.32 Schematic view of the hydrogen bonding between base edges and the benzimidazole groups, in the crystal structures of the Hoechst 33258 molecule complexed with duplex sequences containing the sequence · · · AATT · · · .

selective and effective analogues, for which there are good correlations between DNA binding affinity and biological efficacy. Dicationic diarylfuran molecules (which are almost isostructural with berenil) have shown particular promise in this regard, and the cyclohexyl derivative (Fig. 5.34) is 100-fold more active than pentamidine itself. A crystallographic study of an oligonucleotide complex with this compound has provided a rationale for its improved DNA-binding activity, and indirectly for its biological superiority.[82] The bulky cyclohexyl groups fit snugly into the minor groove, and make extensive

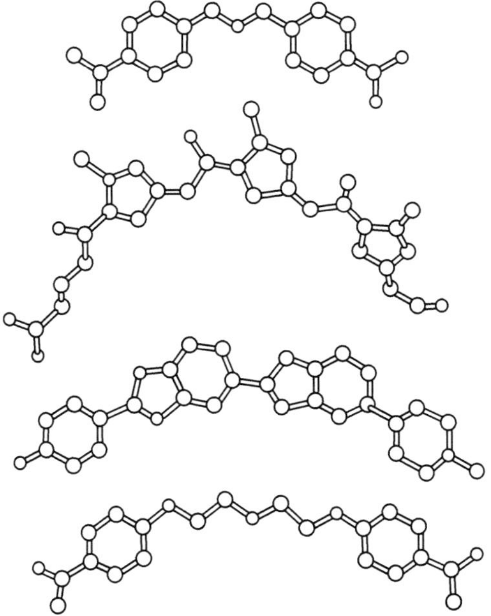

Fig. 5.33 The structures of four minor-groove drugs, taken from the crystal structures of their oligonucleotide complexes. In each case their concave surface is apparent; the degree of concavity differing from one drug to the other, with distamycin having the greatest curvature.

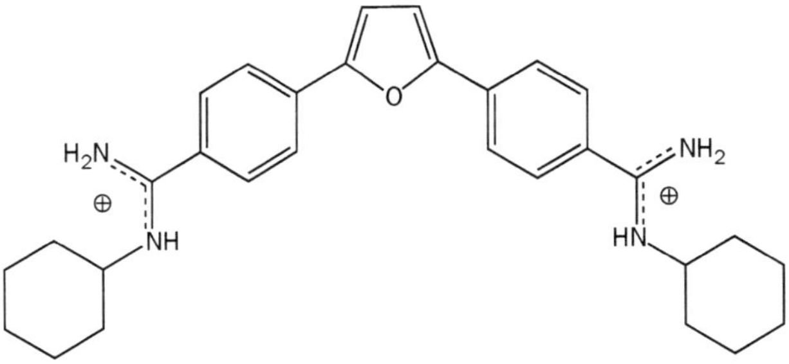

Fig. 5.34 The structure of the cyclohexyl derivative of bis-[amidino-phenyl] furan.

contacts with the non-polar atoms lining the groove walls, to a greater extent than analogues with smaller attached groups.

Hoechst 33258 is widely used as a DNA and chromosomal stain. Footprinting studies have shown that its 4–5 base pair binding site, although necessarily mostly A/T-containing, does have a G/C pair at one end. Structural studies[83–86] have provided an explanation; the two benzimidazole groups in molecule form a network of bifurcated hydrogen bonds to three consecutive A•T base pairs (Figs 5.31 and 5.32). However, the piperidine ring is non-planar and is too bulky to fit into the narrow A/T groove region. Instead, it forces the molecule as a whole into a site with a widened groove at the 3′ end, which is best accommodated by a G•C base pair. The head-to-tail arrangement of the two benzimidazole units in Hoechst 33258 has been extended to three such units, as in the molecule shown in Fig. 5.35. This extended ligand (with the trivial name TRIBIZ) binds tightly, to a site of 7.5 base pairs (Fig. 5.36). As in the parent Hoechst 22358, each benzimidazole group

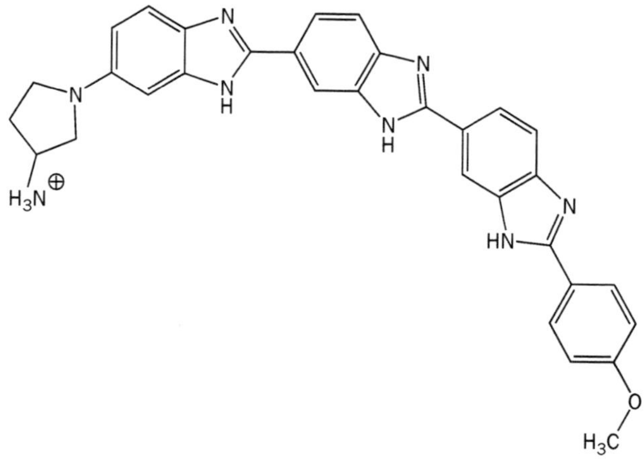

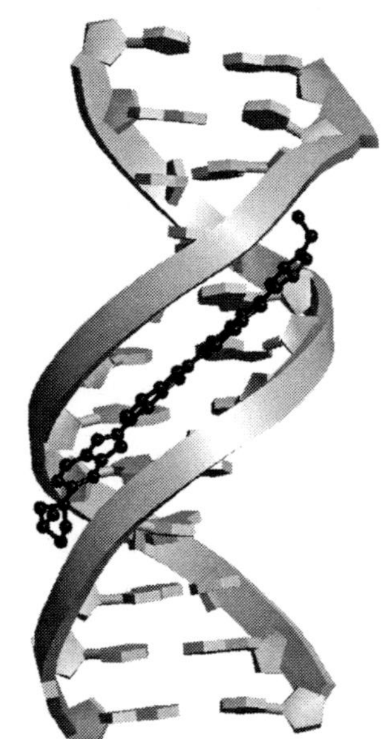

Fig. 5.35 The structure of the groove-binding molecule TRIBIZ, containing three linked benzimidazole units.[109]

hydrogen bonds to two consecutive A•T base pairs by means of a pair of bifurcated hydrogen bonds. The TRIBIZ molecule is just about able to maintain all base pairs in hydrogen-bonding register, with some changes in local DNA structure being needed to achieve this.[87] Further benzimidazole units linked in the same manner would not maintain the correct phasing of hydrogen bonds to successive DNA base pairs. This is a general problem, of 'keeping in register' with successive base pairs in a DNA sequence. It is increasingly significant with ligands designed to recognize upwards of a complete turn of double helix since even small differences in DNA structure and flexibility will be magnified over longer lengths of DNA.

Fig. 5.36 A view of the crystal structure of the complex between the TRIBIZ molecule and the duplex formed by d(CGCAAATTTGCG). The methoxyphenyl group of TRIBIZ is at the upper end of the binding site in this view.

5.6.2 **Netropsin and distamycin**

These two naturally occuring antibiotics (Fig. 5.37) are well-characterized groove-binding drugs, with their DNA interactions studied by a range of biophysical methods.[73] Both comprise linked N-methyl pyrrole and amide units, with netropsin having two cationic charges compared to the one of the larger distamycin molecule. Crystal structures of both drugs complexed with A/T-containing oligonucleotides show that their narrow cross-sections complement the narrow cross-section of the minor groove, again providing an explanation for their observed A/T preferences. Each amide group has its nitrogen atom oriented into the groove and these participate in hydrogen bonding with donor atoms on the edges of adenine and/or thymine bases.[77]

All of these groove binders show a consistent pattern of interaction with A/T stretches of duplex DNA, with two principal sets of observations dominating the sequence selectivity: shape complementarity with the narrow groove walls, and hydrogen bonding to base edges at the groove floor. Implied also, is a negative factor, that the absence of guanine exocyclic $-NH_2$ substituents on the floor of the groove ensures that the shape and hydrogen bonding complementarity are optimal at A/T sites. This concept has been used

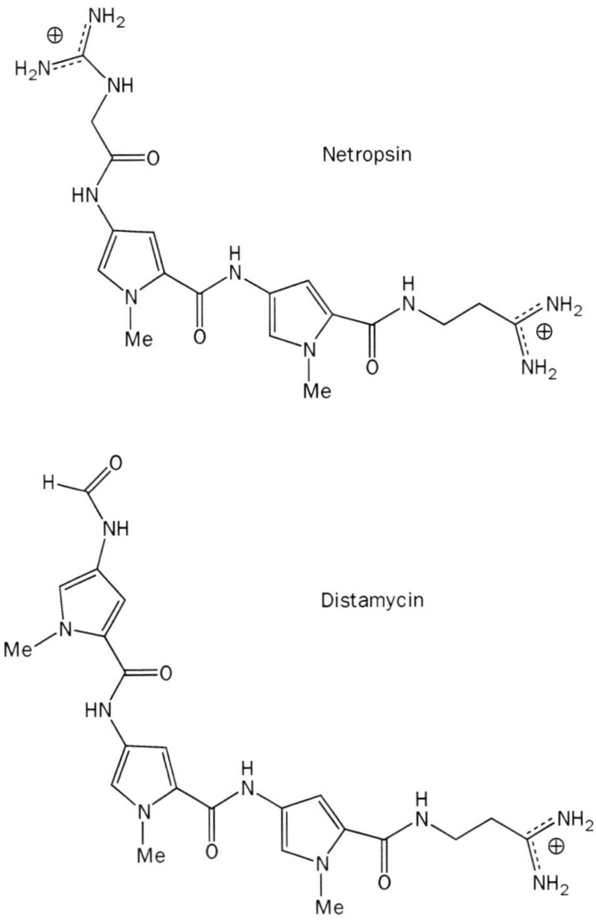

Netropsin

Distamycin

Fig. 5.37 The structures of the oligopeptide-like groove binding drugs netropsin and distamycin.

in the design of molecules with potential G/C selectivity at particular points (Fig. 5.38). Such molecules, termed 'lexitropsins', are based on the proposal[88] of switching hydrogen-bond polarity at the groove floor by means of, for example, an imidazole ring in order to hydrogen-bond to the guanine $-NH_2$ substituent. In practice, lexitropsins have only rarely completely achieved the goal of a total change in sequence specificity. Instead most bind to both A/T and mixed sequences,[89] and tend to show reduced affinities for both types of site. In retrospect this failure can be ascribed to the differences in both groove width and flexibility in G/C compared to A/T regions. The latter tend to be wider and a typical groove-binding molecule with a narrow cross-section will not bind well. The astute reader will also notice that the idealized hydrogen-bonding scheme for lexitropsins implies subtle differences in position for hydrogen-bond donors and acceptors, which are difficult to take into account in the design process. Despite these problems, the lexitropsin concept has proved its value when combined with dimeric groove binding, as discussed below.

An observation, initially from NMR studies,[90] that at higher drug : DNA ratios, distamycin can form a 2 : 1 complex with the duplex sequence d(CGCAAATTTGCG)

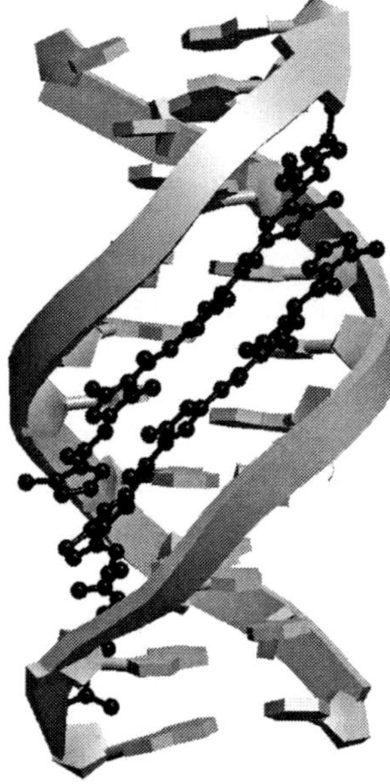

Fig. 5.38 (a) The principles of netropsin and distamycin amide—base recognition, showing hydrogen bonding to the thymine of an A•T base pair; (b) shows the concept of lexitropsin base recognition, with hydrogen bonding from the exocyclic amino substituent of a guanine to the nitrogen atom of an imidazole ring.

Fig. 5.39 The crystal structure[90] of a 2 : 1 complex between distamycin (with atoms coloured black) and the duplex formed by the sequence d(ICICICIC).

(Table 5.4), has been of major significance for subsequent sequence-selective ligand design. This NMR-derived structure has been characterized in detail, and crystallographic analyses have been reported for a number of other 2 : 1 complexes with distamycin bound to a range of sequences (e.g. ref. 91). All show the drug bound in the minor groove as a dimer, in an antiparallel head-to-tail manner (Figs 5.39 and 5.40), in striking contrast to the pattern of 1 : 1 complexes described above, and in particular to the structure of the 1 : 1 distamycin–d(CGCAAATTTGCG) complex. The positively charged ends of the two distamycin molecules are far apart in the dimer complex, suggesting that this arrangement could occur with other groove-binding molecules possessing a single positive charge, such as Hoechst 33258; as yet no such observations have been reported. The NMR and crystallographic structures of the 2 : 1 complexes have both distamycin molecules involved in close non-bonded contacts with each other, and most importantly, each interacts with just one DNA strand (Fig. 5.40). In the 1 : 1 complex of distamycin (and those of netropsin and Hoechst 33258), hydrogen bonding to base edges involves sets of three-centre bifurcated hydrogen bonds between an amide group and any two of adenine N3 and thymine O2 atoms. In order for both distamycin molecules to be accommodated, the 'narrow minor groove' involving the 5′-AAATTT tract must expand from 3.4 to 6.8 Å. Thus, a new picture emerges, so that rather than the minor groove

Table 5.4 Crystal structures of drug–oligonucleotide minor groove complexes

Drug	Sequence	NDB ID code
Distamycin	d(CGCAAATTTGCG)	GDL003
Netropsin	d(CGCGAATTCGCG)	GDBL05
Hoechst 33258	d(CGCGAATTCGCG)	GDL010, GDL011
Hoechst 33258	d(CGCAAATTTGCG)	GDL026, GDL028
Berenil	d(CGCGAATTCGCG)	GDL009
Pentamidine	d(CGCGAATTCGCG)	GDL015
Furamidine	d(CGCGAATTCGCG)	GDL036
Cyclohexyl-furamidine	d(CGCGAATTCGCG)	DD0035
Meta-OH-Hoechst	d(CGCGAATTCGCG)	GDL047, GDL048
TRIBIZ	d(CGCAAATTTGCG)	GDL039
DAPI	d(CGCGAATTCGCG)	GDL008
Mono-imidazole lexitropsin	d(CGCGAATTCGCG))	GLD037, GLD038
Di-imidazole lexitropsin	d(CATGGCCATG)	GDJ054
2 : 1 Distamycin	d(ICICICIC)	GDHB25
Polyamide Im-Hp-Py-Py	d(CCAGTACTGG)	BDD002
Polyamide Im-Py-Hp-Py	d(CCAGATCTGG)	DD0020

having inherently immutable geometry, we have to consider it as being inherently flexible, especially on ligand binding.[86]

5.6.3 Sequence-specific polyamides

The 2 : 1 mode of distamycin binding is the basis for a large number of subsequent studies which have made attainable the goal of sequence recognition at the gene level.[92,93] The key feature of the 2 : 1 complexes is that for the first time minor groove recognition can involve simultaneous hydrogen bonding to both bases in a base pair. Thus, far greater discrimination is inherently attainable than can be achieved with 1 : 1 minor groove binders. A general approach for switching from A : T to G : C base pair recognition has been to adopt a lexitropsin-type of change, whereby one or more N-methyl pyrrole (Py) groups is changed to an imidazole (Im) one (Figs 5.38b and 5.41).

A general recognition code has been established from these studies which enables a wide range of sequences to be recognized using strings of these Im and Py groups linked by amide groups (Table 5.5). The resulting molecules, termed polyamides, have been shown by footprinting analysis to be able to target as dimers a wide range of DNA sequences, with

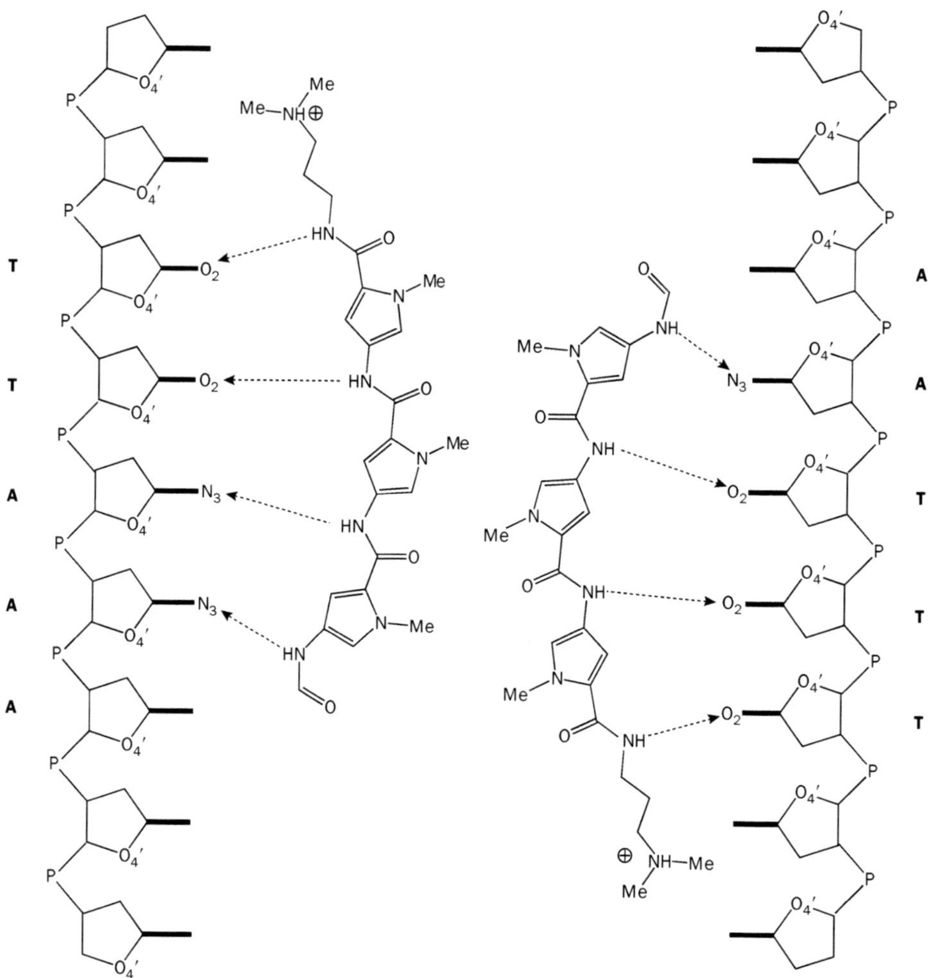

Fig. 5.40 Detail of the hydrogen bonding in the 2 : 1 distamycin dimer complex, showing the hydrogen bonding to eight bases in both DNA strands of a five base pair A/T sequence.

remarkably high selectivity, with typical specific base interactions as shown in Fig. 5.41. Other small heterocyclic rings such as pyrrole (Py) and 3-hydroxypyrrole (Hp) have been successfully incorporated into polyamides to provide hydrogen bonding to particular bases. Greater control of recognition has been achieved by covalently linking two polyamide molecules, which need not have the same sequence of recognition units (Fig. 5.42). This avoids the possibility of slippage between the two unconnected components of a dimer, with potential ambiguity in sequence readout. These hairpin polyamides can achieve exceptionally high selectivity and site affinities, with typical binding constants in the nM range. For example, the molecule Im-Py-Py-Py-γ-Py-Py-Py-Py-β-Dp (where γ is the hairpin linker γ-aminobutyric acid, β is β-alanine and Dp is dimethylaminopropylamide), binds to the sequence d(TGTTAT) with a dissociation constant of 1.1 nM. Several NMR and crystal structures[94–97] have been determined for polyamide– oligonucleotide complexes, which

Table 5.5 The pairing codes for minor-groove polyamide recognition

Pair	C : G	G : C	T : A	A : T
Im/Py	−	+	−	−
Py/Im	+	−	−	−
Hp/Py	−	−	+	−
Py/Hp	−	−	−	+

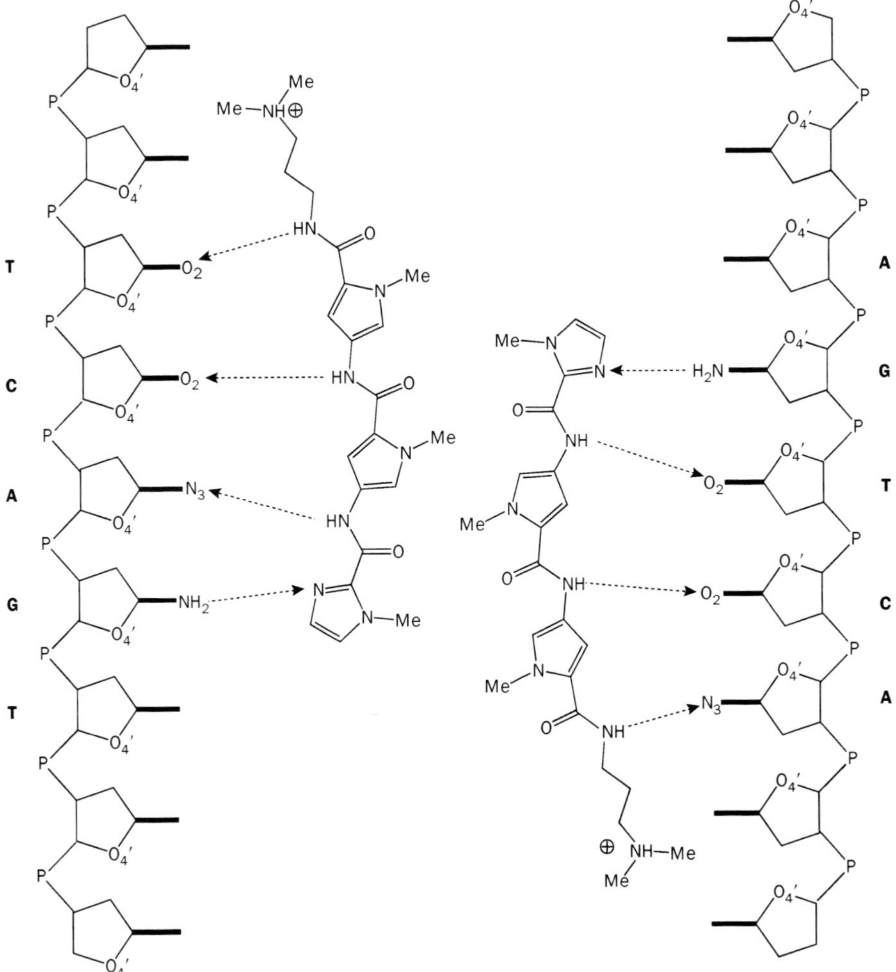

Fig. 5.41 A schematic view of the hydrogen bonding to A•T and G•C base pair edges between the polyamide Im-netropsin and the sequence 5′-TCAGT.

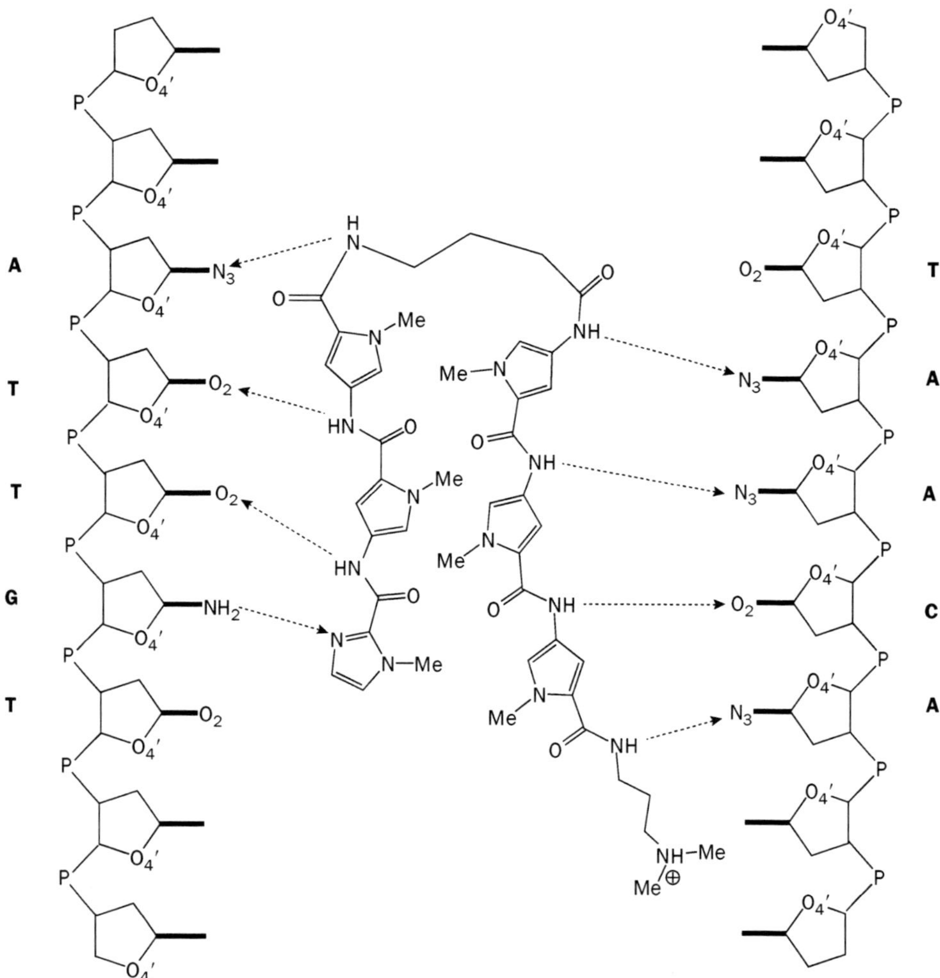

Fig. 5.42 A schematic view of a hairpin polyamide, Im-Py-Py-linker-Py-Py-Py, showing the hydrogen bonding to bases in both strands of a G/C-containing sequence.

have found the predicted patterns of polyamide– base recognition and thus have demonstrated the overall correctness of the approach (Fig. 5.43).

A series of studies have shown that polyamides can target DNA response elements which are sites for transcription factor binding, and successfully compete with these regulatory proteins.[98–100] For example, repression of the 5S RNA gene has been achieved by targeting an eight-ring polyamide to the binding site d(AGTACT) within the TFIIIA transcription factor binding region. The polyamide binds 30-fold more tightly than does TFIIIA to its complete 50-base pair sequence, and effective inhibition of transcription both *in vitro* and in cells was observed. Polyamides can also be used to up-regulate transcription, by blocking repressor sites. A quite unexpected effect has been found[101–103] with a polyamide targeted to the sequence d(GAGAAGAGAA) in the fruit fly *Drosophila*.

This sequence is in satellite DNA, and is normally considered to be functionally inert. It was found that the flies had highly specific loss of function in several defined genes, which are presumed to lie close to and to be affected by binding to satellite DNA. Overall these remarkable effects have been attributed to the experimental observations of drug-induced opening-up of satellite chromatin and a consequent long-range effect on specific genes. It is tempting to speculate that analogous events are involved in the biological effects of the classic A/T-binding minor groove agents.

Are polyamides going to deliver the holy grail of single gene discrimination within a genome? It is probably premature to be able to definitively answer this, since some major obstacles remain before *any* desired 16–18 base pair can be specifically recognized. For one, it is clear that the strategy does not work equally well for all sequences, and that beyond about a length of about 7–9 base pairs, the problem of maintaining polyamide subunits in register sometimes becomes severe. This may be due to differences in flexibility between DNA sequences. Most importantly for therapeutic purposes, polyamides are relatively poorly transported into cells compared to conventional minor-groove drugs such as Hoechst 33258 or pentamidine. A desirable goal is to devise pharmacophore-like building-blocks to mimic polyamides, but as yet, this work

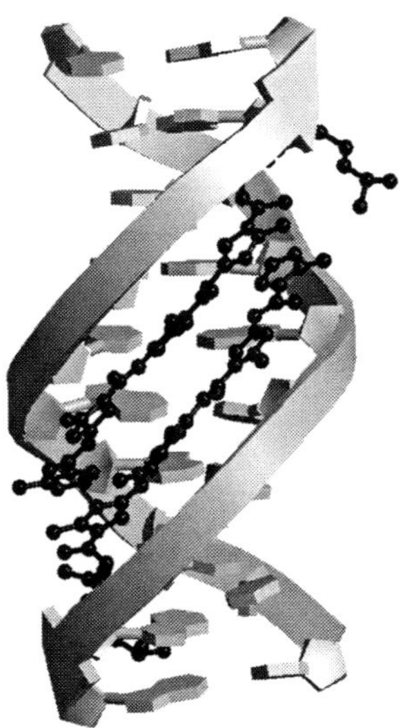

Fig. 5.43 The crystal structure[97] of the polyamide molecule Im-Py-Hp-Py bound to a duplex formed by the sequence d(CCAGATCTGG). The central part of this sequence, that is, 5′-AGATCT, is recognized by this polyamide.

is in its infancy. The repertoire of suitable molecules has been recently extended by the finding[104] that an asymmetric aromatic dication based on the diphenyl furan type of molecule,[82] can also bind in the minor groove as a dimer.

5.7 **Covalent bonding**

The first DNA-interactive anticancer drugs to be developed were alkylating agents, derived from the First World War poison gases. These drugs, the nitrogen mustards, are highly toxic molecules, although their more modern analogues chlorambucil, L-phenylalanine mustard (melphalan) and cyclophosphamide still find use against some cancers. Non-specific covalent binding can take place to the phosphodiester backbone or sugar residues. This is often a step in DNA strand scission, as in the case of the bleomycin family of anticancer drugs. Both single and double-strand breaks can then occur. The latter are usually lethal events to a cell since they are difficult to repair. There has been particular emphasis on studies of binding to particular sites on DNA bases by drugs that undertake nucleophilic reactions. Purines are the most susceptible to covalent attack, with guanine being preferred over adenine. Particularly favoured sites are O6 (guanine), N6 (adenine) and N7 in the major groove and N1, N2 (guanine) and N3 in the minor groove. These site preferences are the result of the differing electronic charge distributions in bases, base pairs and in runs of sequence. The propensity for alkylation by nitrogen mustards at N7 of a guanine base embedded within guanine-rich sequences,[105] is an inherent property of these sequences, since the same pattern of sequence selectivity has been found both *in vitro* and in DNA extracted from nitrogen mustard-treated cells.

Drugs can bind to a single site or to two at once if they have bi-functional capability. This latter cross-linking mode, can be either intra- or inter-strand, depending on two principal factors:

- the distance between the two functional groups on the drug
- the nature of the affected DNA sequence—for example, whether two adjacent guanine are on the same or opposite strands.

Information on the structures of covalent adducts has to date been obtained from on the one hand, chemical and biophysical probe experiments, and on the other, from NMR and some X-ray crystallographic studies. Crystallization of covalent adducts has proved remarkably difficult, and significant success has only been achieved with platinum complexes. In accordance with the theme of this book, we shall focus here on some representative drugs where firm structural information is available, and which are of particular interest at the present time. More extensive accounts of these and the large number of other drugs in this category are available in the reading list at the end of the chapter.

5.7.1 **The platinum drugs**

The chance discovery and subsequent clinical exploitation of the cytotoxic and antitumour properties of the strikingly simple molecule *cis*-dichlorodiamminoplatinum (II) (cisplatin or *cis*-DDP), is one of the great success stories of cancer chemotherapy. It is a highly toxic drug, yet it and its closely related analogues are the probably most effective single agents in current use in the cancer clinic, where it is curative in over 90 per cent of testicular cancers, a disease previously hard to treat, and with a poor prognosis.

Cisplatin mostly binds to purines, especially guanine at the N7 position. A variety of adducts are formed with DNA in solution, with the major species having intrastrand cross-links (Fig. 5.44) to the sequence d(GG), and to a lesser extent to d(AG). A number of more minor species have been characterized, but they are probably of little functional significance. Several NMR and crystal structures have been determined for oligonucleotides containing cisplatin bound at d(GG) sites (Table 5.6). The structure of a

Table 5.6 Oligonucleotide crystal structures with covalently-bound drug molecules

Sequence	Drug	NDB ID code
d(CCTCTG*GTCTCC) + d(GGAGACCAGAGG)	cis-platinum	DDLB73
d(CCTCG*CTCTC) + d(GAGAG*CGAGG)	cis-platinum	DDJ075
d(CCAACGTTGG)	anthramycin	GDJB29

Fig. 5.44 A schematic view of the covalent intrastrand cross-linking interactions of the drug cis-platinum with the N7 atoms of two consecutive guanine bases.

platinum-containing adduct d(CCTCTG⋆CTCTCC)• (d(GGAGACCAGAGG), where G⋆G represents the adduct, has been solved by both techniques,[106,107] and provides an interesting comparison of results from solution versus the crystal. The length of the platinum–N7 bond, of *c.* 2.0 Å, forces the two guanines out of their normal coplanarity, and the conformations of the two adjacent guanosines become highly distorted from native B-DNA. These distortions are not confined to this region of the sequence. The duplex becomes significantly bent, with a large roll towards the major groove at the cisplatin binding site. The NMR study finds a roll of 49° whereas that from the X-ray analysis (Fig. 5.45) is less, of 26°. As a consequence the major groove in both structures becomes more compact and the minor groove is wider and shallower than in canonical B-DNA. The latter changes by up to 6 Å in both width and depth. All these features are more exaggerated in the NMR structure, possibly because of the lack of crystal packing constraints.

These large-scale structural distortions in DNA are remarkably similar to those produced in DNA by the HMG (high-mobility group) chromosomal proteins such as SRY and LEF-1, which are known to bind to the minor groove of DNA. The crystal structure has now been solved of a ternary complex between a 16-base pair duplex containing a single platinum intrastrand cross-link, and an 89-amino acid domain of the HMG1 protein.[108] This is one of the very few drug–DNA–protein ternary complex structures to be determined. It provides a detailed picture of the interplay between the

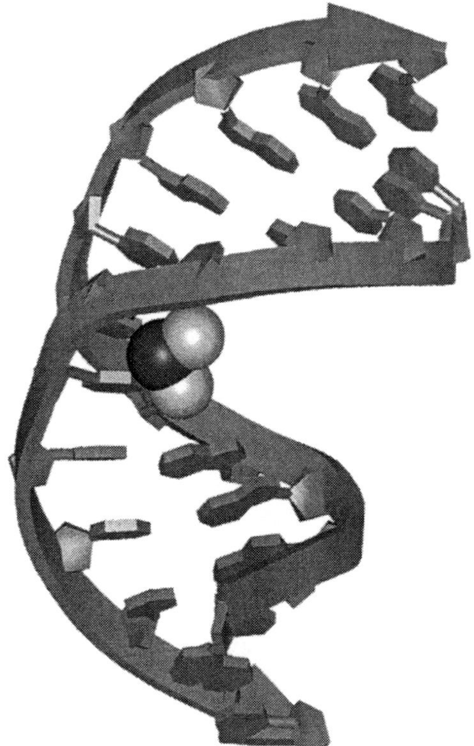

Fig. 5.45 The crystal structure[106] of the covalent complex with cisplatinum formed by the sequence d(CCTCTG⋆GTCTCC)•d(GGAG⋆ ACCA⋆GAGG). G⋆G represents the 1,2-intrastrand adduct. The platinum atom is shown coloured black. Note the marked bend of the two halves of the duplex.

platinum-induced and protein-induced structural changes in DNA, and how platinum lesions in DNA might be recognized by repair proteins. As in the binary platinum adduct, the DNA is sharply bent towards the major groove, but this time by 61°, and the minor groove in this region resembles that in A-form DNA. The protein binding site is wholly on the 3' side of the bound platinum, and the bend itself, whereas the protein–DNA complex alone has protein bound in the centre of the bend. There is more than a passing resemblance between the overall features of the structure and those of the TATA-box–DNA complex.

In striking contrast, the solution structure of DNA containing the less abundant G : G interstrand cross-link has been reported to show local unwinding at the lesion such that there is a reversal at this point to left-handedness and an extra-helical arrangement for the cytosines paired to the cross-linked guanines.[109] This places the platinum in the minor groove. The crystal structure[110] of the same inter-strand cross-link, but with a different sequence (the decamer d(CCTCG*CTCTC)•(d(GAGAG*CGAGG) also shows these features as well as equivalent minor groove widths, although other aspects such as the degree of unwinding at the cross-link and the degree of bending are strikingly different. This suggests that sequence context may be important. However, these structural differences may also reflect the real differences between NMR and X-ray methods in their ability in being able to define long-range structural features with precision. In general, as discussed earlier, NMR methods are not as powerful in defining features such as unwinding and bending when the number of nOe distances is not large.

An even greater degree of distortion has been observed in a 1,3 G : G intrastrand cross-linked solution structure,[111] with a loss of base pairing around the lesion. Again, a pyrimidine base (here a thymine) is extruded from the helix. These consistent observations of extra-helical bases in platinum adducts suggest that they may be a general phenomenon for certain types of platinum lesion. They may act as signals for subsequent DNA repair proteins, suggesting that these minor adducts are relatively unimportant for the anti-tumour activity of cisplatin compared to the 1,2-intrastrand lesion, which is less readily repaired and, therefore, more persistent.

5.7.2 Covalent binding combined with sequence-specific recognition

Sequence-specific DNA recognition is well illustrated in the case of the anti-tumour antibiotic (+)-CC-1065 (Fig. 5.46), which has three pyrrolo-indole units, and its simpler analogue duocarmycin, with just one. In both instances, one end of the drug molecule binds covalently to N3 of an adenine via opening of the cyclopropane ring.[112] The rest of the drug lies in the minor groove. CC-1065 covers four base pairs to the 5' side of this adenine and one base pair on the 3' side. Specific recognition is to sequences such as 5'-AAAAA; the drug induces bending of 17–22°, comparable to that found in natural A tracts, which enables a close steric fit between drug and DNA to take place. So CC-1065 has a specific (and highly stringent) requirement for a sequence that will bend in an appropriate manner. The smaller drug duocarmycin induces fewer structural changes in its target sequences, and the overall conformations remain close to canonical B-DNA.[113] Bizelesin is a synthetic analogue of CC-1065 with high experimental anti-tumour activity. It has a chemically reactive chloromethyl group at each end of the molecule, and was rationally designed to covalently cross-link to the hexamer sequence

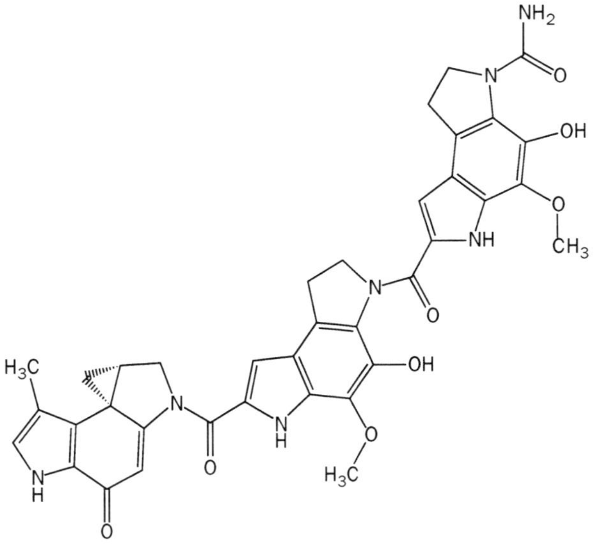

Fig. 5.46 The structure of the covalently binding drug CC-1065.

5'-TAATTA. This sequence is intrinsically bent in solution, as a consequence of the high propeller twist of the A·T base pairs. Unexpectedly, NMR studies[114] have shown that there are two adduct conformations in a 40 : 60 ratio. Both have straightened-out A tracts, but differ in the hydrogen-bonding arrangement of the central A·T base pairs, with one being unpaired and the other having a Hoogsteen arrangement.

Minor-groove alkylation most often occurs at the N2 of guanine. A number of compounds that bind here have significant experimental anti-tumour activity. Mitomycin, discovered in Japan in fermentation broths almost 50 years ago, has found some clinical application in the treatment of solid tumours, although its severe toxicity has precluded extensive use. Several other minor-groove alkylators are currently being evaluated for clinical trials. Their biological efficacy is undoubtedly a consequence of interference with particular aspects of gene regulation, although this is as yet poorly understood. For at least some compounds such as DSB-120[115] and its analogues (see below), their binding to the minor groove of B-form DNA involves little, if any, perturbation from the canonical structure. This results in a lack of structural lesions for subsequent recognition by a repair protein, so that these adducts tend to be long-lasting, with consequent enhancement of potency.

The pyrrolo[2,1-c][1,4]benzodiazepine (PBD) family of compounds, typified by the naturally occurring anti-tumour antibiotic anthramycin, bind to the exocyclic N2 atom of guanine bases through the N10–C11 double bond (Fig. 5.47). The crystal structure[116] of an adduct with two bound drug molecules (Fig. 5.48) shows that each lies in the minor groove, in a manner analogous to that of the non-covalent binders such as berenil and netropsin. The binding site for anthramycin is three bases long, with a preference for 5'-Pu-G-Pu type sequences. Several dimers of these molecules have been designed which have exceptional cytotoxicity. The best-studied of these is DSB-120,[115] with an IC_{50} of 0.0005 μM, making it some 1000 times more potent than the equivalent monomer in

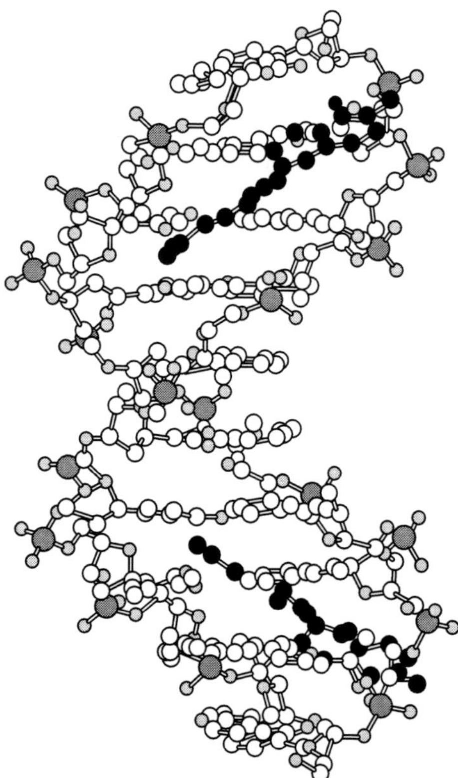

Fig. 5.47 The drug anthramycin (top), showing its point of attachment to the exocyclic amino group of guanine. The anthramycin dimer DSB-120 is shown below.

Fig. 5.48 The crystal structure[116] of a complex with two anthramycin molecules bound to the duplex formed by the sequence d(CCAACGTTGᴬG). Gᴬ represents the adduct attachment point.

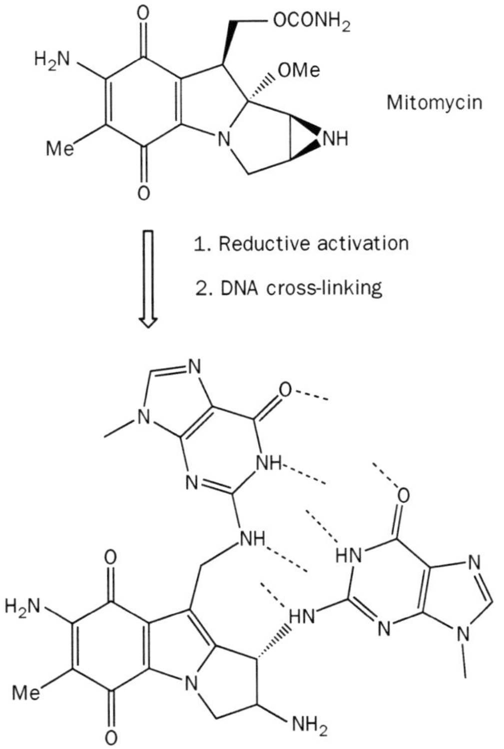

Fig. 5.49 Structure of mitomycin, together with that of the DNA adduct cross-linked to two guanines on opposite strands of a DNA duplex.

some cell lines. These cytotoxic effects are due to sequence-selective inhibition of transcription. DSB-120 (Fig. 5.47) has a site size requirement of 6–7 base pairs, with a sequence preference for sites of the type 5′-Pu-GATC-Py. Its ability to form inter-strand covalent cross-links to two sites in the minor groove, together with its almost ideal isohelicity, result in a tight fit to canonical B-form DNA. NMR and molecular modelling studies have shown that DSB-120 produces almost no structural perturbation to DNA upon cross-linking. This feature of the PDB dimer complexes is undoubtedly significant for their resistance to DNA repair and may contribute to the high *in vivo* anti-tumour activity shown by several members of this series.

Mitomycin C requires metabolic activation in order to react with DNA.[117] Opening of the aziridine ring (Fig. 5.49) results in covalent attachment to one guanine in the sequence 5′-CpG. A second is through the carbamate group. The structures of both this cross-linked adduct and a mono-adduct from aziridine ring-opening, have been characterized by NMR methods,[118,119] as has the very different N7 adduct, which places the bound drug in the DNA major groove,[120] without significant perturbations from B-DNA form.

High sequence selectivity is a dominant feature of several of these minor-groove covalent binding molecules. This, together with their drug-like characteristics, does suggest that a fruitful future direction for the development of therapeutic sequence-selectivity at the gene level, may be in combining such molecules with the features of non-covalent recognition, such as have been described earlier in this chapter.

References

1   International Human Genome Sequencing Consortium, (2001). *Nature*, **409**, 860.

2   Seeman, N. C., Rosenberg, J. M., and Rich, A. (1976). *Proceedings of the National Academy of Sciences USA*, **73**, 804.

3   Pullman, A. and Pullman, B. (1981). *Quarterly Reviews of Biophysics*, **14**, 289.

4   Kuninec, M. G. and Wemmer, D. E. (1992). *Journal of the American Chemical Society*, **114**, 8739.

5   Liepinsh, E., Otting, G., and Wüthrich, K. (1992). *Nucleic Acids Research*, **20**, 6549.

6   Phan, A. T., Leroy, J.-L., and Guéron, M. (1999). *Journal of Molecular Biology*, **286**, 505.

7   Feig, M. and Pettitt, B. M. (1999). *Journal of Molecular Biology*, **286**, 1075.

8   Duan, Y., Wilkosz, P., Crowley, M., and Rosenberg, J. M. (1997). *Journal of Molecular Biology*, **272**, 553.

9   Schneider, B. and Berman, H. M. (1995). *Biophysical Journal*, **69**, 2661.

10   Schneider, B., Patel, K., and Berman, H. M. (1998). *Biophysical Journal*, **75**, 2422.

11   Alden, C. J. and Kim, S.-H. (1979). *Journal of Molecular Biology*, **132**, 411.

12   Scheider, B., Cohen, D., and Berman, H. M. (1992). *Biopolymers*, **32**, 725.

13   Saenger, W., Hunter, W. N., and Kennard, O. (1986). *Nature*, **324**, 385.

14   Kennard, O., Cruse, W. B. T., Nachman, J., Prange, T., Shakked, Z., and Rabinovich, D. (1986). *Journal of Biomolecular Structure and Dynamics*, **3**, 623.

15   Drew, H. R. and Dickerson, R. E. (1981). *Journal of Molecular Biology*, **151**, 535.

16   Shui, X., McFail-Isom, L., Hu, G. G., and Williams, L. D. (1998). *Biochemistry*, **37**, 8341.

17   Tereshko, V., Minasov, G., and Egli, M. (1999). *Journal of the American Chemical Society*, **121**, 470.

18   Chiu, T. K., Kaczor-Grzeskowiak, M., and Dickerson, R. E. (1999). *Journal of Molecular Biology*, **292**, 589.

19   Liu, J. and Subirana, J. (1999). *Journal of Biological Chemistry*, **274**, 247.

20   Minasov, G. Tereshko, V., and Egli, M. (1999). *Journal of Molecular Biology*, **291**, 83.

21   Johansson, E., Parkinson, G., and Neidle, S. (2000). *Journal of Molecular Biology*, **300**, 551.

22   Shui, X., Sines, C. C., McFail-Isom, L., Van Derveer, D., and Williams, L. D. (1998). *Biochemistry*, **37**, 16877.

23   Hud, N. V., Sklenár, V., and Feigon, J. (1999). *Journal of Molecular Biology*, **286**, 651.

24   Billeter, M., Güntert, P., Luginbühl, P., and Wüthrich, K. (1996). *Cell*, **85**, 1057.

25   McConnell, K. J. and Beveridge, D. L. (2000). *Journal of Molecular Biology*, **304**, 803.

26   Lerman, L. S. (1961). *Journal of Molecular Biology*, **3**, 18.

27   Waring, M. J. (1970). *Journal of Molecular Biology*, **54**, 247.

28   Neidle, S., Achari, A., Taylor, G.L., Berman, H. M., Carrell, H. L., Glusker, J. P., and Stallings, W. C. (1977). *Nature*, **269**, 304.

29   Wang, A. H.-J., Ughetto, G., Quigley, G. J., and Rich, A. (1987). *Biochemistry*, **26**, 1152.

30   Kamitori, S. and Takusagawa, F. (1992). *Journal of Molecular Biology*, **225**, 445.

31   Liu, X., Chen, H., and Patel, D. J. (1991). *Journal of Biomolecular NMR*, **1**, 323.

32   Egli, M., Williams, L. D., Frederick, C. A., and Rich, A. (1991). *Biochemistry*, **30**, 1364.

33   Williams, H. E. L. and Searle, M. S. (1999). *Journal of Molecular Biology*, **290**, 699.

34   Fox, K. R. and Waring, M. J. (1986). *Biochemistry*, **25**, 4349.

35   Todd, A. K., Adams, A., Thorpe, J. H., Denny, W. A., Wakelin, L. P. G., and Cardin, C. J. (1999). *Journal of Medicinal Chemistry*, **42**, 536.

36   Adams, A., Guss, J. M., Collyer, C. A., Denny, W. A., Prakash, A. S., and Wakelin, L. P. G. (2000). *Molecular Pharmacology*, **58**, 649.

37   Kielkopf, C. L., Erkkila, K. E., Hudson, B. P., Barton, J. K., and Rees, D. C. (2000). *Nature Structural Biology*, **7**, 117.

38   Adams, A., Guss, J. M., Collyer, C. A., Denny, W. A., and Wakelin, L. P. G. (2000). *Nucleic Acids Research*, **28**, 4244.

39   Thorpe, J. H., Hobbs, J. R., Todd, A. K., Denny, W. A., Charlton, P., and Cardin, C. J. (2000). *Biochemistry*, **39**, 15055.

40   Yang, X.-L., Robinson, H., Gao, Y.-G., and Wang, A. H.-J. (2000). *Biochemistry*, **39**, 10950.

41   Eichman, B. F., Mooers, B. H. M., Alberti, M., Hearst, J. E., and Ho, P. S. (2001). *Journal of Molecular Biology*, **308**, 15.

42   Marzilli, L. G., Banville, D. L., Zon, G., and Wilson, W. D. (1986). *Journal of the American Chemical Society*, **108**, 4188.

43   Lipscomb, L. A., Zhou, F. X., Presnell, S. R., Woo, R. J., Peek, M. E., Plaskon, R. R., and Williams, L. D. (1996). *Biochemistry*, **35**, 2818.

44   Bennett, M., Krah, A., Wien, F., Garman, E., McKenna, R., Sanderson, M., and Neidle, S. (2000). *Proceedings of the National Academy of Sciences USA*, **97**, 9476.

45   Wang, A. H.-J., Ughetto, G., Quigley, G. J., Hakoshima, T., van der Marel, G. A., van Boom, J. H., and Rich, A. (1984). *Science*, **225**, 1115.

46   Quigley, G. J., Ughetto, G., van der Marel, G. A., van Boom, J. H., Wang, A. H.-J., and Rich, A. (1986). *Science*, **232**, 1255.

47   Gao, X. and Patel, D. J. (1988). *Biochemistry*, **27**, 1744.

48   Gilbert, D. E. and Feigon, J. (1992). *Nucleic Acids Research*, **20**, 2411.

49   McLean, M. J., Seela, F., and Waring, M. J. (1989). *Proceedings of the National Academy of Sciences USA*, **86**, 9687.

50   Chaires, J. B., Leng, F., Przewloka, T., Fokt, I., Ling, Y.-H., Perez-Soler, R., and Priebe, W. (1997). *Journal of Medicinal Chemistry*, **40**, 261.

51   Hu, G. G., Shui, X., Leng, F., Priebe, W., Chaires, J. B., and Williams, L. D. (1997). *Biochemistry*, **36**, 5940.

52   Peek, M. E., Lipscomb, L. A., Bertrand, J. A., Gao, Q., Roques, B. P., Garbay-Jaureguiberry, C., and Williams, L. D. (1994). *Biochemistry*, **33**, 3794.

53   Mergny, J. L., Duval-Valentin, G., Nguyen, C. H., Perrouault, L., Faucon, B., Rougée, Montenay-Garestier, T., *et al.* (1992). *Science*, **256**, 1681.

54   Cassidy, S. A., Strekowski, L., Wilson, W. D., and Fox, K. R. (1994). *Biochemistry*, **33**, 15338–15347.

55   Fox, K. R., Polucci, P., Jenkins, T. C., and Neidle, S. (1995). *Proceedings of the National Academy of Sciences USA*, **92**, 7887.

56   Silver, G. C., Sun, J.-S., Nguyen, C. H., Boutorine, A. S., Bisagni, E., and Hélène, C. (1997). *Journal of the American Chemical Society*, **119**, 263.

57   Escude, C., Nguyen, C. H., Kukreti, S., Janin, Y., Sun, J. S., Bisagni, E., *et al.* (1998). *Proceedings of the National Academy of Sciences USA*, **95**, 3591.

58   Guo, Q., Lu, M., Marky, L. A., and Kallenbach, N. R. (1992). *Biochemistry*, **31**, 2451.

59   Mergny, J. L. and Hélène, C. (1998). *Nature Medicine*, **4**, 1366.

60   Han, H. and Hurley, L. H. (2000). *Trends in Pharmaceutical Sciences*, **21**, 136.

61   McEachern, M. J. Krauskopf, A., and Blackburn, E. H. (2000). *Annual Reviews in Genetics*, **34**, 331.

62   Cech, T. R. (2000). *Angewante Chemie International Edition*, **39**, 34.

63   Zahler, A. M, Williamson, J. R., Cech, T. R., and Prescott, D. M (1999). *Nature*, **350**, 718.

64   Sun, D., Thompson, B., Cathers, B. E., Salazar, M., Kerwin, S. M., Trent, J. O., *et al.* (1997). *Journal of Medicinal Chemistry*, **40**, 2113.

65   Wheelhouse, R. T., Sun, D., Han, H., Han F. X., and Hurley, L. H. (1998). *Journal of the American Chemical Society*, **120**, 3261.

66   Read, M. A., Wood, A. A., Harrison, J. R., Gowan, S. M., Kelland, L. R., Dosanjh, H. S., and Neidle. S. (1999). *Journal of Medicinal Chemistry*, **42**, 4538.

67   Koeppel, F., Riou, J.-F., Laoui, A., Mailliet, P., Arimondo, P. B., Labit, D., *et al.* (2001). *Nucleic Acids Research*, **29**, 1087.

68   Mergny, J.-L, Lacroix, L., Teulado-Fichou, M.-P., Hounsou, C., Guittat, L., Hoarau, M., *et al.* (2001). *Proceedings of the National Academy of Sciences USA*, **98**, 3062.

69   Read, M., Harrison, R. J., Romagnoli, B., Tanious, F. A., Gowan, S. H., Reszka, A. P., *et al.* (2001). *Proceedings of the National Academy of Sciences USA*, **98**, 4844.

70   Fedoroff, O. Y., Salazar, M., Han, H., Chemeris, V. V., Kerwin, S. M., and Hurley, L. H. (1998). *Biochemistry*, **37**(36), 1236.

71   Read, M. A. and Neidle, S. (2000). *Biochemistry*, **39**, 13422.

72   Rangan, A., Fedoroff, O. Y., and Hurley, L. H. (2001). *Journal of Biological Chemistry*, **276**, 4640.

73   Zimmer, C. H. and Wähnert, U. (1986). *Progress in Biophysics and Molecular Biology*, **47**, 31.

74   Dervan, P. B. (1986). *Science*, **232**, 464.

75   Brown, D. G., Sanderson, M. R., Garman, E., and Neidle, S. (1992). *Journal of Molecular Biology*, **226**, 481.

76   Lane, A. N., Jenkins, T. C., Brown, T., and Neidle, S. (1991). *Biochemistry*, **30**, 1372.

77   Kopka, M. L., Yoon, C., Goodsell, D., Pjura, P., and Dickerson, R. E. (1986). *Journal of Molecular Biology*, **183**, 553.

78   Patel, D. J. and Shapiro, L. (1986). *Journal of Biological Chemistry*, **261**, 1230.

79   Goodsell, D. and Dickerson, R. E. (1986). *Journal of Medicinal Chemistry*, **29**, 727.

80   Abu-Daya, A., Brown, P. M., and Fox, K. R. (1995). *Nucleic Acids Research*, **17**, 3385.

81   Laughton, C. and Luisi, B. (1999). *Journal of Molecular Biology*, **288**, 953.

82   Boykin, D. W., Kumar, A., Spychala, J., Zhou, M., Lombardy, R. J., Wilson, W. D., *et al.* (1995). *Journal of Medicinal Chemistry*, **38**, 912.

83   Vega, M. C., García Sáez, I., Aymami, J., Eritja, R., Van der Marel, G. A., Van Boom, J. H., *et al.* (1994). *European Journal of Biochemistry*, **222**, 721.

84   Spink, N., Brown, D. G., Skelly, J. V., and Neidle, S. (1994). *Nucleic Acids Research*, **22**, 1607.

85   Gavathiostis, E., Sharman, G. J., and Searle, M. S. (2000). *Nucleic Acids Research*, **28**(3), 728–735.

86   Bostock-Smith, C. E., Laughton, C. A., and Searle, M. S. (1999). *Biochemical Journal*, **342**, 125.

87   Clark, G. R., Gray, E. J., Neidle, S., Ji, Y.-H., and Leupin, W. (1996). *Biochemistry*, **35**, 13745; Ji, Y.-H., Bur, D., Häsler, W., Schmitt, V. R., Dorn, A., Bailly, C., *et al.* (2001). *Bioorganic and Medicinal Chemistry*, **9**, 2905.

88   Lown, J. W. (1994). *Journal of Molecular Recognition*, 7, 79.

89   Kopka, M. L., Goodsell, D. S., Han, G. W., Chiu, T. K., Lown, J. W., and Dickerson, R. E. (1997). *Structure*, **5**, 1033.

90   Pelton, J. G. and Wemmer, D. E. (1990). *Journal of the American Chemical Society*, **112**, 1393.

91   Chen, X., Ramakrishnan, B., Rao, S. T., and Sundaralingam, M. (1994). *Nature Structural Biology*, **1**, 169.

92   Mrksich, M. and Dervan, P. B. (1993). *Journal of the American Chemical Society*, **115**, 2572.

93   Dervan, P. B. and Bürli, R. W. (1999). *Current Opinion in Structural Biology*, **3**, 688.

94   de Clairac, R. P. L., Geierstanger, B. H., Mrksich, M., Dervan, P. B., and Wemmer, D. E. (1997). *Journal of the American Chemical Society*, **119**, 7909.

95   Kielkopf, C. L., Baird, E. E., Dervan, P. B., and Rees, D. C. (1998). *Nature Structural Biology*, **5**, 104.

96   Kielkopf, C. L., White, S., Szewczyk, J. W., Turner, J. M., Baird, E. E., Dervan, P. B., and Rees, D. C. (1998). *Science*, **282**, 111.

97   Kielkopf, C. L., Bremer, R. E., White, S., Szewczyk, J. W., Turner, J. M., Baird, E. E., *et al.* (2000). *Journal of Molecular Biology*, **295**, 557.

98   Gottesfeld, J. M., Neely, L., Trauger, J. W., Baird, E. E., and Dervan, P. B. (1997). *Nature*, **387**, 202; Gottesfield, J. M., Melander, C., Suto, R. K., Raviol, H., Luger, K., and Dervan, P. B. (2001). *Journal of Molecular Biology*, **309**, 615.

99   Dickinson, L. A., Trauger, J. W., Eldon, E. B., Ghazal, P., Dervan, P. B., and Gottesfeld, J. M. (1999). *Biochemistry*, **38**, 10801.

100 Mapp, A. K., Ansari, A. Z., Ptashne, M., and Dervan, P. B. (2000). *Proceedings of the National Academy of Sciences USA*, **97**, 3930; Bremer, R. E., Wurtz, N. R., Szewczyk, J. W., and Dervan, P. B. (2001). *Bioorganic and Medicinal Chemistry*, **9**, 2093.

101 Henikoff, S. and Vermaak, D. (2000). *Cell*, **103**, 695.

102 Janssen, S., Durussel, T., and Laemmli, U. K. (2000). *Molecular Cell*, **6**, 999.

103 Janssen, S., Cuvier, O., Müller, M., and Laemmli, U. K. (2000). *Molecular Cell*, **6**, 1013; Maeshima, K., Janssen, S., and Laemmli, U. K. (2001). The *EMBO Journal*, **20**, 3218.

104   Wang, L., Bailly, C., Kumar, A., Ding, D., Bajic, M., Boykin, D. W., and Wilson, W. D. (2000). *Proceedings of the National Academy of Sciences USA*, **97**, 12.

105   Hartley, J. A., Bingham, J. P., and Souhami, R. L. (1992). *Nucleic Acids Research*, **20**, 3175.

106   Takahara, P. M., Frederick, C. A., and Lippard, S. J. (1996). *Journal of the American Chemical Society*, **118**, 12309.

107   Gelasco, A. and Lippard, S. J. (1998). *Biochemistry*, **37**, 9230.

108   Ohndorf, U.-M., Rould, M. A., He, Q., Pabo, C. O., and Lippard, S. J. (1999). *Nature*, **399**, 708.

109   Huang, H., Zhu, L., Reid, B. R., Drobny, G. P., and Hopkins, P. B. (1995). *Science*, **270**, 1842.

110   Coste, F., Malinge, J.-M., Serre, L., Shepard, W., Roth, M., Leng, M., and Zelwer, C. (1999). *Nucleic Acids Research*, **27**, 1837.

111   Teuben, J.-M., Bauer, C., Wang, A. H.-J., and Reedijk, J. (1999). *Biochemistry*, **38**, 12305.

112   Hurley, L. H., Lee, C.-S, McGovren, J. P., Mitchell, M. A., Warpehoski, M. A., Kelly, R. C., and Aristoff, P. A. (1988). *Biochemistry*, **27**, 3886.

113   Eis, P. S., Smith, J .A., Rydzewski, J. M., Chase, D. A., Boger, D. L., and Chazin, W. J. (1997). *Journal of Molecular Biology*, **272**, 237.

114   Thompson, A. S. and Hurley, L. H. (1995). *Journal of Molecular Biology*, **252**, 86.

115   Bose, D. S., Thompson, A. S., Ching, J., Hartley, J. A., Berardini, M. D., Jenkins, T. C., *et al.* (1992). *Journal of the American Chemical Society*, **114**, 4939.

116   Kopka, M. L., Goodsell, D .S., Baikalov, I., Grzeskowiak, K., Cascio, D., and Dickerson, R. E. (1994). *Biochemistry*, **33**, 13593.

117   Tomasz, M and Palom, Y. (1997). *Pharmacology and Therapeutics*, **76**, 73.

118   Sastry, M., Fiala, R., Lipman, R., Tomasz, M., and Patel, D. J. (1995). *Journal of Molecular Biology*, **247**, 338.

119   Norman, D., Live, D., Sastry, M., Lipman, R., Hingerty, B. E., Tomasz, M., *et al.* (1990). *Biochemistry*, **29**, 2861.

120   Subramaniam, G., Paz, M. M., Kumar, G. S., Das, A., Palom, Y., Clement, C. C., *et al.* (2001). *Biochemistry*, **40**, 10473.

Further reading

DNA–water interactions

Berman, H. M. (1991). *Current Opinion in Structural Biology*, **1**, 423.

Berman, H. M. (1994). *Current Opinion in Structural Biology*, **4**, 345.

Westhof, E. (1988). *Annual Reviews in Biophysics*, **17**, 125.

Westhof, E. and Beveridge, D. L. (1990). In *Water Science Reviews 5.* (ed. Franks, F.) Cambridge University Press, Cambridge, UK.

DNA–drug interactions

Denison, C. and Kodadek, T. (1998). *Chemistry & Biology*, **5**, 129.

Dervan, P. B. (2001). *Bioorganic and Medicinal Chemistry*, **9**, 2215.

Gale, E. F., Cundliffe, E. F., Reynolds, P. E., Richmond, M. H., and Waring, M. J. (1981). *The Molecular Basis of Antibiotic Action.* John Wiley, London.

Gao, X. and Patel, D. J. (1989). *Quarterly Reviews in Biophysics*, **22**, 93.

Hurley, L. H. (1989). *Journal of Medicinal Chemistry*, **30**, 2027.

Neidle, S. (2001). *Natural Product Reports*, **18**, 291.

Neidle, S. and Waring, M. J. (eds) (1993). *Molecular Basis of Drug-DNA Antitumour Action.* Macmillan Press, London. Vols 1, 2.

Pratt, W. B., Ruddon, R. W., Ensminger, W. D., and Maybaum, J. (1994). *The Anticancer Drugs*, 2nd edn, Oxford University Press, New York.

Rajski, S. R. and Williams, R. M. (1998) *Chemical Reviews*, **98**, 2723.

Reddy, B. S. P., Sondhi, S. M., and Lown, J. W. (1999). *Pharmacology and Therapeutics*, **84**, 1.

Ren, J. and Chaires, J. B. (1999). *Biochemistry*, **38**, 16067.

Thurston, D. E. (1999). *British Journal of Cancer*, **80**, 65.

Wemmer, D. E. and Dervan, P. B. (1997). *Current Opinion in Structural Biology*, 7, 355.

Yang, X.-L. and Wang, A. H.-J. (1999). *Pharmacology and Therapeutics*, **83**, 181.

6

The RNA structural world

6.1 Introduction

The preceding chapters have illustrated the structural variability displayed by DNA. Virtually all of this is within the boundaries imposed by the over-riding structural requirements of double, triple or four-stranded helices. RNA molecules are not subject to these limitations, and consequently their structural variety is in striking contrast to the relative predictability of most DNA tertiary structure. It is perhaps unsurprising (at least in retrospect!) that the ability of many RNAs to fold into compact tertiary structures, analogous to the folding of polypeptides into proteins and enzymes, is paralleled in some instances, by the possession of catalytic activity.

The need for an RNA intermediary between the code for a gene contained in DNA, and the ultimate gene product, a protein, was realized soon after the discovery of the structure of DNA itself. This category of (single-stranded) RNA molecule was initially conceived of as a carrier of the code to the ribosome, although this picture of RNA as an intermediary is somewhat complicated by the involvement of double-stranded RNA in RNA viruses. The need for a second type of RNA molecule, transfer RNA (tRNA), was also recognized early in the history of molecular biology, in order to perform the specialized function of decoding and transferring the information contained in an individual codon in a messenger RNA molecule into the correct amino acid. An organism has 20 sets of tRNA molecules, each set specific for a particular amino acid (in principle only one tRNA is required for each amino acid, but in practice there are more than this, with some redundancy being needed for several amino acids). A specific tRNA attaches a correct amino acid and transfers it to a specialized region of the ribosome, where it is incorporated into a growing protein chain. More recently, a third yet more specialized function for RNA has been found, that of RNA enzymatic activity, for catalytic cleavage of RNA (and DNA) itself. Such cleavage can be intramolecular, such as self-splicing introns in RNA editing. RNA cleavage by RNA molecules, termed ribozymes, has excited much interest with the suggestion that they could be potential key players in a prebiotic pre-DNA world. Ribozymes are also being studies for their potential therapeutic use in gene therapy, for cleavage and ultimate destruction of specific DNA sequences.

RNA crystallographic and NMR structural studies have advanced rapidly over the past few years. There have been several technical advances that have made this possible. On the one hand, solid-phase RNA chemical synthesis and purification (of sequences up to 15–20 bases in length) is now usually relatively straightforward for the preparation of milligram quantities, although the methods are still not as straightforward as those for DNA oligomer preparation.[1,2] Even short RNA sequences are very prone to secondary structure formation, and special care in purification is needed. Larger sequences are routinely prepared enzymatically, using efficient *in vitro* transcription of the appropriate DNA sequence with T7 RNA polymerase, although there is a requirement for a purine sequence at the 5' end in order for optimal efficiency of transcription to take place. Also, random sequence nucleotides are sometimes added at the 3' end by T7 polymerase. These problems can be overcome by methods that can post-transcriptionally excise the desired sequence.[3] Commercial kits are now available for RNA transcription, which can readily produce milligram quantities of sequences up to several hundred bases in length. Crystallization of RNAs has traditionally been difficult, in part on account of their inherent flexibility. The introduction of sparse matrix methods coupled with robotic technology has enabled success to be achieved relatively rapidly with a number of RNAs.[4–6] These methods enable a very large range of crystallization conditions to be rapidly sampled, using only small quantities of RNA. However, success by screening alone is never guaranteed, and methods have been successfully developed which rationally engineer specific intermolecular contacts in order to optimize the formation of crystalline arrays of molecules.[7] Crystal structures of the majority of small, purely helical RNAs are still routinely solved by molecular replacement methods, normally using canonical duplex RNA models as starting points. Synthetic RNAs, as with DNAs, can incorporate heavy atoms such as bromine for MAD phasing, on specific residues. These are needed when molecular replacement does not work, or with more complex structures.[8] Large, complex RNAs always require heavy-atom derivatives, and systematic approaches have been developed for rapidly homing in on those heavy atom reagents that are most likely to produce useable derivatives, such as osmium hexamine. A useful strategy in some instances is to use a small synthetic RNA sequence containing a heavy-atom derivative, which can become an integral part of the total RNA molecule when annealed with the larger RNA component prepared by transcription.

6.2 **Fundamentals of RNA structure**

6.2.1 **Helical RNA conformations**

RNA differs from DNA in two key chemical aspects (Fig. 6.1):

(1) the sugar groups in all RNAs have an OH group attached at the 2' position, making them ribose rather than deoxyribose sugars;

(2) thymine bases are replaced by uracils, which have the same hydrogen-bonding, but do not have a methyl group at the 5 position.

The effect of the extra hydroxyl group is profound. RNA sugars are much more rigid than DNA ones, and only adopt the C3'-*endo* pucker type (Fig. 6.2) since otherwise the

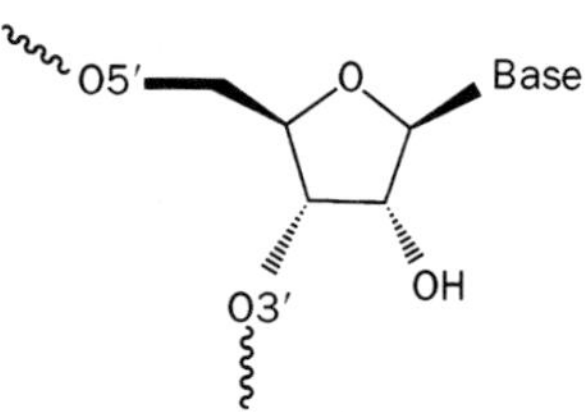

Fig. 6.1 Configuration of a ribose sugar in RNA.

2'-OH group would clash with C8 (for a purine) or C6 (for a pyrimidine) of the attached base. Simulation studies, together with analysis of crystal structures, have shown that this pucker is stabilized in several positive ways.[9] The 2'-OH group is oriented only in several discrete positions as a result of interactions with O3' or O4', or with accessible hydrogen-bond acceptors on the attached base. In helical A-RNA, the important interaction is between the H2' atom and O4' of the next residue, which is not accessible for hydrogen-bonding to water molecules.

Although single-stranded RNAs do not adopt rigid conformations in the absence of restraining tertiary interactions (see below), double-stranded RNA can be readily formed, analogous to duplex DNA. Uracil participates in U•A base pairs that are fully isomorphous with A•T ones in duplex DNA. Duplex RNA is conformationally rather rigid, and its behaviour contrasts remarkably with the polymorphism of duplex DNA in that only one major polymorph of the RNA double helix has been observed. This has many features in common with A-DNA, and accordingly is known as A-RNA. The conformational features of canonical RNA helices have been obtained from fibre diffraction studies on duplex RNA polynucleotides from both viral and synthetic origins. A-RNA is an 11-fold helix, with a narrow and deep major groove and a wide, shallow minor groove, and base pairs inclined to and displaced from the helix axis (Fig. 6.3a,b). It is notable that the base pairs themselves in canonical A-RNA are only slightly twisted. The major A-RNA form is apparent in natural polynucleotide fibres, or in those formed from poly(A)•poly(T) in conditions of low ionic strength.[10] At higher ionic strength, a slightly different 12-fold helix (A'-RNA) is formed, which retains the same overall helical features, although its wider major groove resembles that of A-DNA. These are detailed in Table 6.1. Both helices have the C3'-*endo* sugar pucker characteristic of ribopolynucleotides. Another difference from duplex DNA is that RNA helices, though capable of a small degree of bending (up to *c.* 15°), does not undergo the large-scale bending seen, for example, in A-tract DNA.

A number of single-crystal and NMR studies on sequences forming perfectly base paired duplex RNA have been reported. These have shown that the RNA geometry observed in fibres is conserved across a wide range of sequences, provided that there are no base mismatches, which can under some circumstances, greatly alter helical structure (see below). As a consequence of their greater

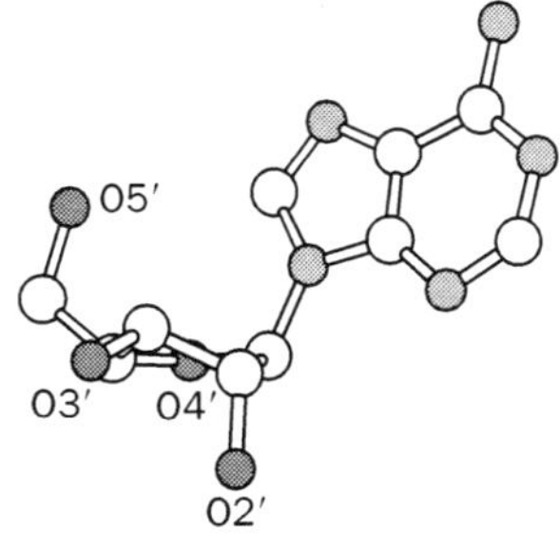

Fig. 6.2 A ribonucleoside in an A-RNA conformation.

(a)

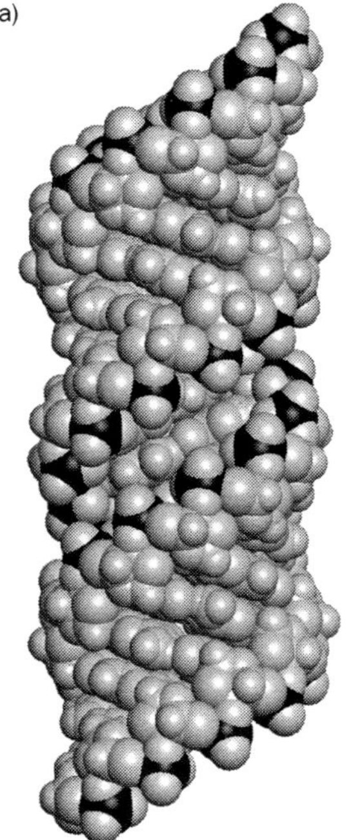

(b)

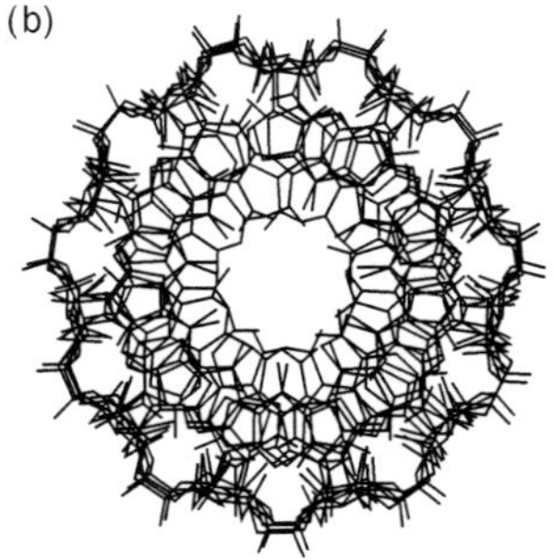

Fig. 6.3 (a) Space-filling representation of an A-RNA double helix; (b) view down the helix axis of an A-RNA double helix, now drawn in stick mode. Note that the base pairs sit astride the helix axis.

Table 6.1 Structural features of RNA double helices, from fibre diffraction studies[11]
(a) Conformational angles. Values for poly(A)•poly(T) in the A form are given

		α	β	χ	δ	ϵ	ξ	χ
A-RNA	Adenine	−69	179	55	82	−154	−71	−161
	Thymine	−64	178	51	83	−152	−173	−161
A′-RNA		−70	177	61	77	−153	−70	−163

(b) Groove dimensions in Å

	Major-groove width	Major-groove depth	Minor-groove width	Minor-groove depth
A-RNA	4.7	12.9	10.8	3.3
A′-RNA	8.9	14.4	10.5	3.4

(c) Selected helical and base/base pair morphological parameters

	Base displacement (Å)	Rise (Å)	Helical twist (°)	Roll (°)	Propeller twist (°)	Inclination (°)
A-RNA	0.44	2.8	32.7	−0.8	−2.1	−15.5
A′-RNA	0.51	3.0	30.0	0.0	2.3	−10.6

conformational rigidity RNA helices do not have the pronounced sequence-dependent microstructure of duplex DNA.

The first reported single-crystal RNA analyses were of the self-complementary dinucleosides r(AU) and r(GC).[12,13] These determinations, at atomic resolution, are of historic importance in showing Watson–Crick base pairing within the context of short helical segments, and in demonstrating that even dinucleoside duplexes have backbone conformations characteristic of helical A-RNA. They provided the impetus for structural studies of intercalative drug binding to dinucleoside duplexes as models for the polynucleotide situation (see Chapter 5).

Crystallographic analyses of sequences such as the octanucleotide r(CCCCGGGG), have shown helices of length sufficient for a full set of helical parameters to be extracted. This sequence crystallizes in two distinct crystal lattices, enabling the effects of crystal packing factors on structure to be assessed.[14] In each instance, rhombohedral and hexagonal, the RNA helices are very similar, and their features closely resemble those in fibre diffraction canonical A-RNA, with, for example, minor-groove widths of 9.8 and 9.7 Å respectively. Visualization of a complete turn of RNA double helix requires a longer sequence. The structure of r(UUAUAUAUAUAUAA) was the first to show this.[15] Again, the helix is essentially classical A-RNA (Fig. 6.4), with an average helical twist of 33° and an axial rise of 2.8 Å. The structure of a mixed sequence, r(UAAGGAGGU-GAU), has been determined,[16] which is a more rigorous test of the conservation of RNA helical structure (Table 6.2). This RNA dodecamer sequence closely resembles a region required for the initiation of translation in prokaryotics. The overall features of the two independent helices in this crystal structure are of A-RNA type rather than A′-RNA, with, for example, helical twists of 33.3° and 33.5° respectively, corresponding to a helical repeat of 10.7–10.8 base pairs/turn. Longer sequences are more difficult to crystallize in ordered conformations. This is illustrated by the analysis of two duplexes, an 18- and a 19-mer, which diffract to high resolution, 1.20 and 1.55 Å respectively.[17] Despite this

Table 6.2 RNA helix crystal structures

Sequence	NDB number
r(CCCCGGGG)	ARH074
r(UAAGGAGGUGAU)	ARL062
r(UUAUAUAUAUAUAA)	ARN035
r(GGGGGGGGGGGG)	ARO020

apparent indication of a high degree of internal order, the 18-mer is disordered such that each base pair is an average of the duplex as a whole. This is closely analogous to the fibre diffraction situation, and occurs in both long DNA and RNA duplexes when the crystal packing involves continuous helices. The 19-mer is disordered in a distinct way, with two orientations being apparent. Both sequences are in classic A-RNA conformations, except for the sugar puckers, which have a significantly larger pucker amplitude than in simple ribonucleosides.

The increasing availability of high-resolution crystal structures has enabled RNA hydration in oligonucleotides to be defined. Some general rules are beginning to emerge for the water arrangements beyond the obvious finding of an inherently greater degree of hydration compared to DNA oligonucleotides, on account of the 2′-OH group being an active and willing hydrogen bond participant. There is no analogue of the minor-groove spine of hydration seen in B-DNA. Again, this is unsurprisingly since in A-RNA the minor groove is too wide for such an arrangement. A detailed analysis[18] of the crystal structure of r(CCCCGGGG) at a resolution of 1.8 Å has shown that the base edges and phosphate groups tend to be more extensively hydrated than in DNA. This appears to be in large part due to the 2′-OH groups, with the 16 in this duplex having 33 hydrogen bonds to water molecules.

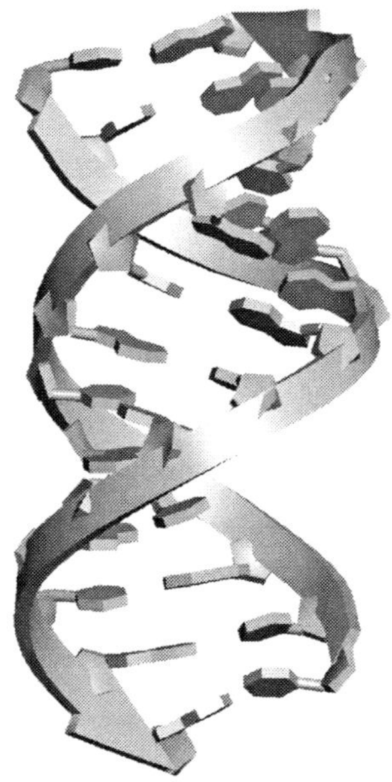

Fig. 6.4 The crystal structure of the duplex of r(UUAUAUAUAUAUAA), shown in schematic form.[15] Note the characteristic A-RNA inclination of base pairs with respect to the vertical helix axis.

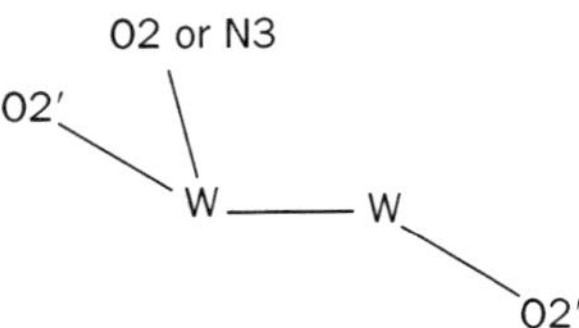

Fig. 6.5 The transverse hydrogen-bonded water linkage across the minor groove in the structure of r(CCCCGGGG).[18]

These 2′-OH act as foci for the establishment of water networks in both major and minor grooves. There is, for example, a set of seven identical well-defined transverse hydrogen-bonded water linkages across the minor groove (Fig. 6.5). Long time-scale molecular dynamics simulations on equivalent DNA and RNA oligomers have confirmed the slightly greater hydration of the latter,[19] with *c.* 21.4 water molecules per RNA C•G base pair compared to 19.8 for a B-DNA one. Counter-ions such as sodium or potassium tend to accumulate in the major groove of RNA duplexes,[19,20] by contrast with their minor-groove localization in B-DNA.

6.2.2 Mismatched and bulged RNA structures

RNAs can readily form stable base pair and triplet mismatches and extrahelical regions within the context of 'normal' RNA double helices. Such features, notably the extrahelical loops (Fig. 6.6), have been the subject of intense structural study, since they are present in large RNAs (tRNAs, mRNAs, ribosomal RNA), and together with various types of base stacking, are responsible for maintaining their tertiary structures. It is paradoxically common for crystal structures of short sequences containing potential loop regions such as the UUCG 'tetraloop' (which forms an especially stable extrahelical loop structure in solution), to often not show such features (Table 6.3). This is probably a consequence of the high ionic strength of many crystallization conditions, together with the preference of some sequences to pack in the crystal as helical arrays. So, instead of loops, these crystal structures tend to have runs of non-Watson–Crick mismatched base pairs where the loop would have formed. For example, an A-RNA double helix, albeit with G•U and U•U base pairs, is formed in the crystal[21] by the sequence r(GCUUCGGC)d(BrU). There is evidence of some deviations from the exact canonical A-RNA duplexes formed by fully Watson–Crick base pairs, since this helix has <10 base pairs per turn. The dodecamer sequence r(GGACUUCGGUCC) similarly forms a base paired duplex,[22] with U•C and G•U base pairs. The A-RNA helix here has a significant increase in the width of its major groove, possibly on account of the water molecules that are strongly associated with the mismatch base pairs.

Much greater perturbations are apparent[23] in the structure of the dodecamer r(GGCCGAAAGGCC), where the four non-Watson–Crick base pairs in the centre of the sequence form an internal loop with sheared G•A and A•A base pairs. The resulting structure is very distorted from A-RNA ideality (Fig. 6.7), with a compression of the major groove, an enlargement of the minor-groove width to 13.5 Å, and a pronounced curvature of the resulting helix. Again, this sequence forms a tetraloop in solution.[24] The G•U base pair is a prominent and very important element of large RNA structures since it is especially stable,[25] on account of its two hydrogen bonds (Fig. 6.8). It is known as the 'wobble' pair, since a G in the first position of a codon can accept either a C or a U in the third anticodon (wobble) position. The RNA genome of the HIV-1 retrovirus (HIV-1 Rev) contains many non-helical features, some of which have

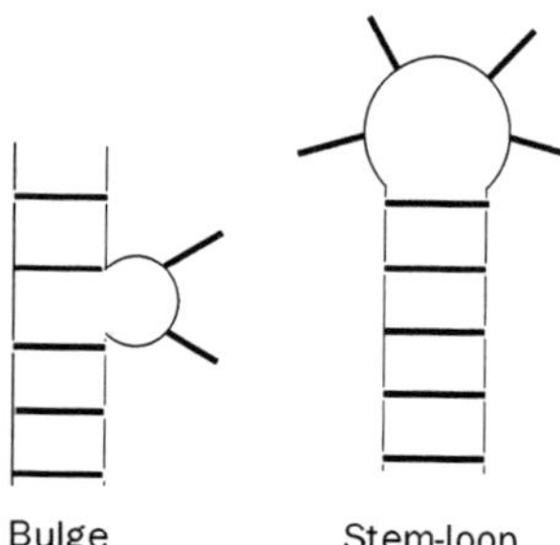

Fig. 6.6 Two types of extra-helical features in RNA.

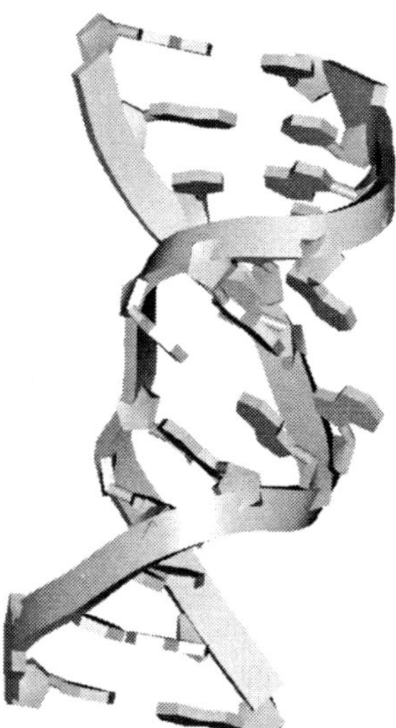

Fig. 6.7 The structure of the dodecamer r(GGCCGAAAGGCC).[23]

Table 6.3 Crystal structures of small RNAs with non-helical features

Sequence	Feature	NDB number
r(GGACUUCGGUCC)	U•C, G•U base pairs	ARL037
r(GGCCGAAAGGCC)	G•A, A•A base pairs	URL051
29-nucleotide SRP	purine bulge	AR0024
r(GCGCGGCACCGUCCG CGGAACAAACGG)	pseudoknot	UR0004
HIV-1 RRE site	G•G, A•G base pairs; A, U extrahelical	AR0023

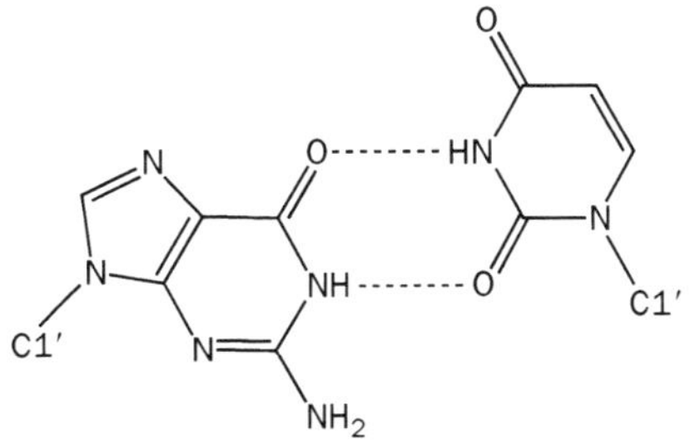

Fig. 6.8 The G•U wobble base pair.

Fig. 6.9 The sequence of the dimerization initiation site of the RNA genome from the HIV-1 retrovirus, studied crystallographically.[26] The helical and extra-helical adenosines are highlighted.

been studied by structural methods. Its dimerization initiation site has features that act as signals for RNA packaging. Again the expected secondary structure (containing two loops in a 'kissing-loop' arrangement) was not observed. Instead, the duplex contains two A•G base pairs (Fig. 6.9), each adjacent to an extra-helical, bulged-out adenosine.[26] This tendency of short RNA sequences to maximize helical features is also apparent in the crystal structure of a 29-nucleotide fragment from the signal recognition particle,[27] which forms 28-mer heteroduplexes rather than a hairpin structure. Even so, this duplex has features of wider interest, since it has a number of non-Watson–Crick base pairs such as a 5′-GGAG/3′-GGAG purine bulge. Their overall effect is to produce backbone distortions so that the helix has non-A-RNA features such as a widening of the major groove, by ~9 Å, and local under-twisting of base pairs adjacent to A•C and G•U mismatches.

NMR approaches to determining RNA structure have been fruitful in defining the conformation of double-helical RNA. However, sometimes the geometries of non-helical regions are less unambiguously defined as a consequence of both their inherent flexibility and the relative paucity of nOe signals leading to a lack of sufficient constraints to unambiguously define the NMR model. The correspondence (and complementarity) between crystallographic and NMR structures is well illustrated in studies of the high-affinity RNA binding site of the HIV-1 Rev protein response element, whose conformation changes significantly with bound protein. The NMR and X-ray structures on the RNA itself agree in describing the RRE as a highly distorted helix.[28,29] The distortions are a consequence of (i) two extra-helical bases, an adenine and a uracil from one strand, which are responsible for a bulge and a kink respectively in the helix, and (ii) two essential purine•purine base pairs, which flank the extra-helical uracil (Fig. 6.10). Both structures show that the helix is bent by ~30°, with a narrowed major groove, and is highly distorted from canonical A-RNA. The principal difference between them is in the loop region around the extra-helical uracil. This is a consequence of the adjacent G(*syn*)•G(*anti*) base pair having an asymmetric configuration in the X-ray structure, whereas in the NMR structure it has been assigned a symmetric one, so that the backbones are forced to adopt distinct paths in this region. Whether these differences are real, or reflect differing solution conditions, remains to be established.

We have noted above that crystallographic studies of small RNAs containing putative loop regions invariably result in helices that instead maximize base pairing. The structure of the structurally important and very stable UUCG tetraloop has been especially elusive until recently. Initial NMR studies proved to be only partially correct. A more recent analysis[30] is in close agreement with a crystallographic analysis of a tetraloop

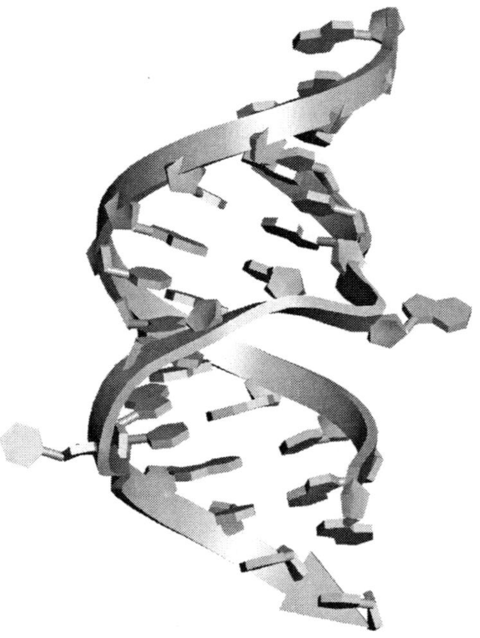

Fig. 6.10 The structure of the
RNA binding site of the HIV-1 Rev
protein response element.[28,29]

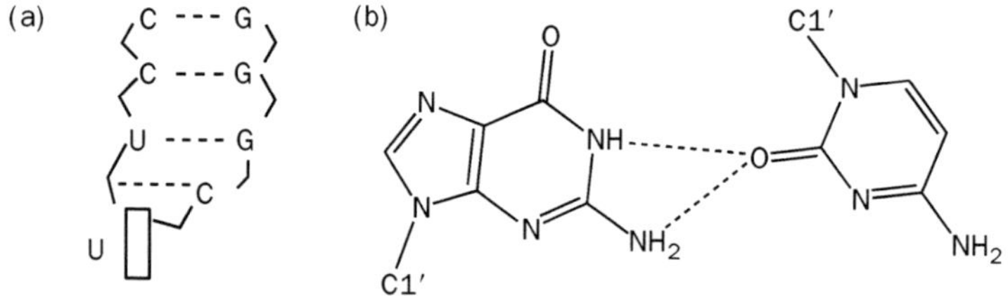

Fig. 6.11 (a) A schematic view of the UUCG tetraloop,[30–32] with the U
base looped out from pairing; (b) the G(*syn*)•U(*anti*) base pair found in
the UUCG tetraloop.[30–32]

incorporated in a large RNA–protein crystal structure,[31] and molecular dynamics simulations,[32] which converged to the correct structure from the incorrect starting-point. The
essential features of the UUCG tetraloop (Fig. 6.11a) are (i) an unusual G(*syn*)•U(*anti*)
base pair (Fig. 6.11b), which is seen directly in the X-ray structure, (ii) the 2′-OH group
of this uridine nucleoside hydrogen bonds to the O6 atom of the guanine, (iii) the cytosine
stacks on the adjacent uracil base.

The relative accuracy of crystallography and NMR is hard to assess unless identical
structures are compared. This has been done for two small RNAs, one of which is fully
helical, and the other has a small loop in it.[33] Gratifyingly, the differences between each
pair of structures are small. The study also compared various parameter sets, commonly
used for nucleic acid X-ray and NMR refinements, and concluded that deficiencies in
them (especially in electrostatics) are probably responsible for many of the (small) differences observed between solution and crystal structures. The overall message is that it

should be the goal of solution structure determination to reproduce as closely as possible the local and global structures found by high-resolution crystallography.

6.3 **Transfer RNA structures**

We have already seen in Chapter 4 that DNA sequences are able to participate in non-Watson–Crick base–base interactions. RNA sequences are inherently even more capable of doing so, and indeed they are the norm in large natural RNA structures, once Watson–Crick base pairing is maximized. There is a significant literature on RNA secondary structure prediction, and a number of computer algorithms are freely available with which one can examine RNA sequences. They are based, in large part, on the maximizing of base pairing. Successful prediction also takes into account the existence of a number of invariant features in a sequence across species. Historically, the determination of the primary sequence of a tRNA molecule in 1964 rapidly led to suggestions for its secondary folding into a cloverleaf structure. This fold is a consequence of base–base hydrogen bonding between non-adjacent regions of RNA double helix and loop. It is based on the identification of the highly conserved features of acceptor, anticodon, D and T arms, and three/four loops, one of which is very variable between tRNAs (the 'variable' loop). There were numerous attempts at predicting the three-dimensional structure of a tRNA. None were completely accurate, which is unsurprising given the very large number of potential base–base interactions. However, the general notion that these base–base interactions are responsible for RNA folding, has proved to be correct, as were the assignments of Watson–Crick base pairs. All tRNAs contain, in addition to the standard A, U, G and C nucleosides, a number of modified nucleosides such as N2-dimethylguanosine, N1-methyladenosine, pseudouridine (Ψ; an isomer of uridine with the glycosidic bond from the ribose sugar going to the C5 atom of the base rather than to N1), and dihydrouridine D. tRNA species have between 70 and 95 nucleosides, depending in large part on the size of the variable loop.

The crystal structure of yeast phenylalanine tRNA was determined in the early 1970s, simultaneously by two groups.[34,35] These together with a few other tRNAs (Table 6.4) remained the sole complex RNA molecular structures available for 20 years, until the first ribozyme structure. Both independent structures, one monoclinic[34] and the other orthorhombic,[35] are closely similar. They showed that the molecule is folded into an overall L-shape, with the two arms at right angles to each other. The original cloverleaf is still apparent, but with additional interactions between distant parts of the structure (Fig. 6.12). The arms consist of short A-RNA helices, together with these extensive base–base interactions that hold the two arms together (Fig. 6.13). The longer double

Table 6.4 Selected tRNA crystal structures

tRNA	NDB number
Yeast Phe, MIT	TRNA04
Yeast Phe, LMB	TRNA03
Yeast Phe, re-refinement	TRNA09
Yeast Phe, high resolution	TR0001
Yeast Asp	TRNA05
Yeast initiator	TRNA12

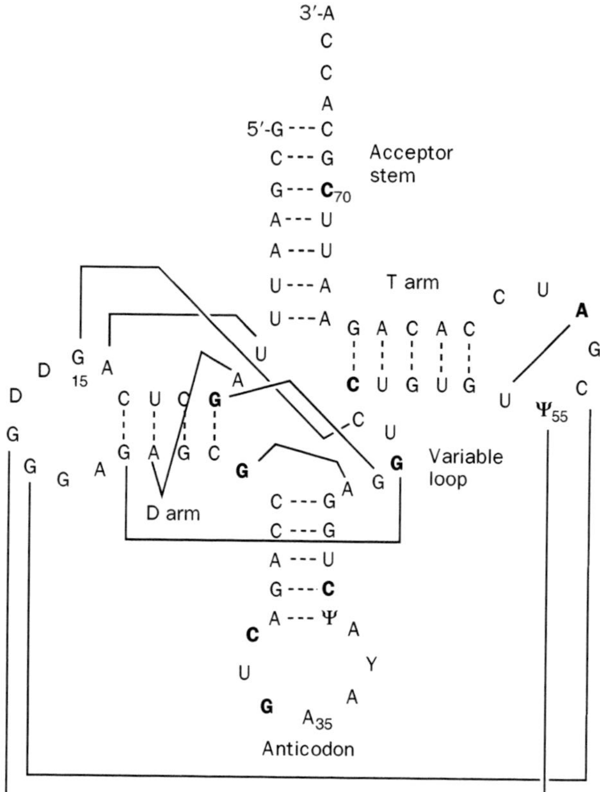

Fig. 6.12 Schematic of the hydrogen-bonding interactions in the yeast tRNA[phe] crystal structure.[34,35] The interactions between nucleotides that appear to be remote in this, the clover-leaf representation.

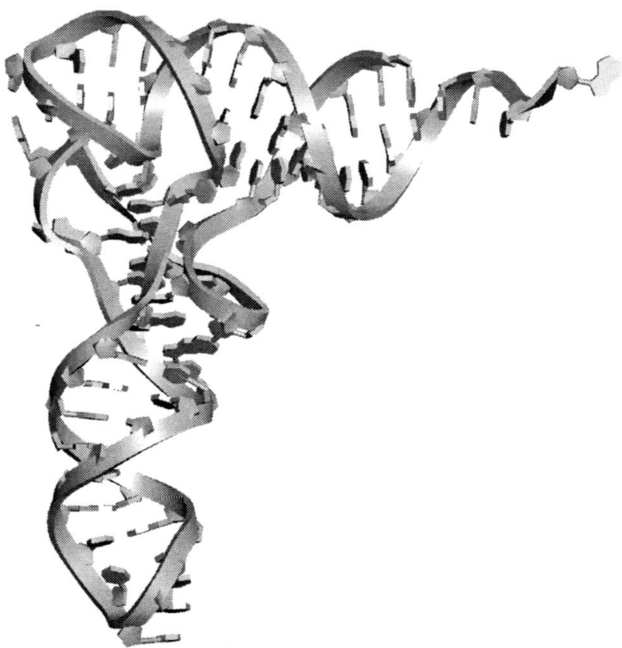

Fig. 6.13 The crystal structure of yeast tRNA[phe], showing the L-shape.[34,35]

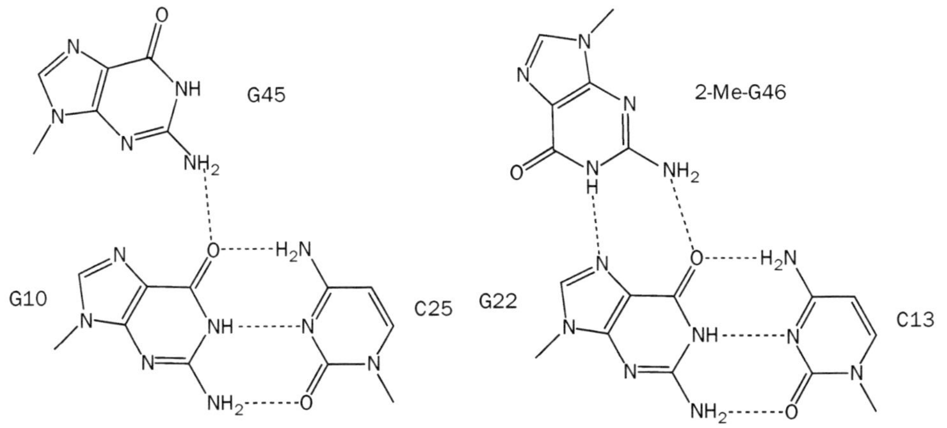

Fig. 6.14 Some base triplet interactions in the yeast tRNA[phe] crystal structure.

helical anticodon arm has the short helix of the D stem stacked upon it. The other arm is formed by the helix of the acceptor stem, on which is stacked the four base pair T arm helix. This key feature of helix–helix stacking has turned out to be of central importance for other complex RNAs. The nine additional base · · · base interactions that maintain the structural fold all tend to be in the elbow region, where the two arms are joined together. Almost all of these are of non-Watson–Crick type, and several are triplet interactions (Fig. 6.14). These are analogous to the triplets found in DNA triple helices (Chapter 4). Other subsequent crystal structures of tRNAs (Table 6.4) have shown that the overall L shape is invariant, as are many of the tertiary interactions.

The yeast tRNA[phe] crystal structures have been re-refined several times subsequent to their initial determinations, each time using increasingly robust refinement methods, so that in effect these structures have been used as a test-bed for the general development of nucleic acid crystal structure refinement methodology.[36,37] Most recently the crystal structure of the monoclinic form has been re-determined[38,39] using synchrotron sources to collect data, to 1.93 and 2.0 Å resolution respectively, from flash-frozen crystals. It is reassuring that the overall backbone geometry and fold remains as found in the earlier studies. There are corrections to the sugar pucker and conformation for a very few nucleotides, and backbone torsion angles conform much more to the canonical A-RNA values—this reflects both the improved quality of the diffraction data and the robustness of current refinement methods. Both of these new analyses find new magnesium binding sites, and a greater number of well-defined water molecules hydrogen-bonding to the tRNA.

6.4 **Ribozymes**

The ability of certain RNA molecules to catalytically cleave either themselves or other RNAs is shown by several distinct categories of RNA molecule, with undoubtedly more remaining to be discovered:

1. The RNA of self-splicing group I introns—that from *Tetrahymena* was the first ribozyme to be discovered.[40,41] These contain four conserved sequence elements and form a characteristic secondary structure. The initial step of the cleavage involves nucleophilic attack by a conserved guanosine.

2. The RNA of self-splicing group II introns. These also have conserved sequence elements, but a very different secondary structure and a distinct mechanism of cleavage that involves the nucleophilic attack of a conserved adenosine, contained within the intron sequence.
3. The RNA subunit of the enzyme RNase P.
4. Self-cleaving RNAs from viral and plant satellite RNAs. These are smaller ribozymes than the group I or II intron ones, and include the hammerhead ribozyme (Fig. 6.15).
5. Ribosomal RNA in the ribosome.

6.4.1 **The hammerhead ribozyme**

The crystal structures of hammerhead ribozymes were the first to be determined, simultaneously by two groups (Table 6.5).[42,43] Both structures are similar in their essential detail. One[42] has a DNA strand containing the putative cleavage site. Since ribozymes do not cleave DNA, this is effectively an inhibitor complex. The other[43] is all-RNA, with a 2′-O-methyl group replacing the 2′-hydroxyl group of the active-site cytosine. The structure shows three A-RNA stems connected to a central two-domain region containing the conserved residues (Fig. 6.16). The overall arrangement resembles a wishbone. The two-residue sequence CA (the site of autocatalytic cleavage) is at the apex of two stems (Fig. 6.15), where there is a sharp turn in backbone conformation, that is focused on the uridine in domain one, formed by the invariant tetranucleotide sequence CUGA. This U-turn conformation is closely similar to those in the anticodon and pseudouridine loops of tRNA. The second domain, at the junction of two stems, contains conserved sheared

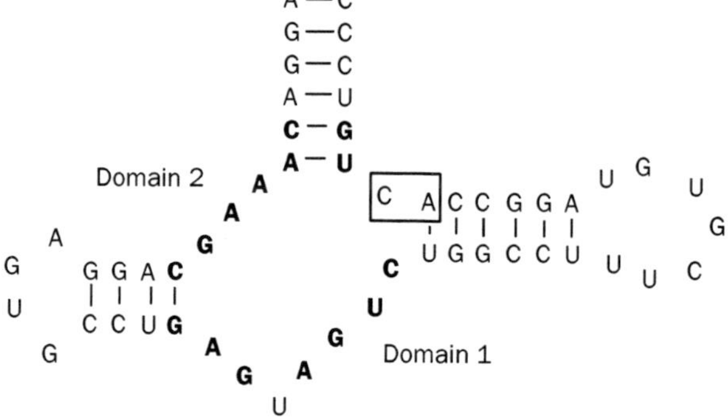

Fig. 6.15 Sequence of a hammerhead ribozyme, indicating the hydrogen-bonded helical stems. The CA cleavage site is shown boxed.

Table 6.5 Ribozyme crystal structures

Ribozyme type	NDB number
RNA Hammerhead	URX057
P4-P6 domain	URX053
Tetrahymena ribozyme	UR0003
Hepatitis delta virus	PR0005
Hairpin	PR0038

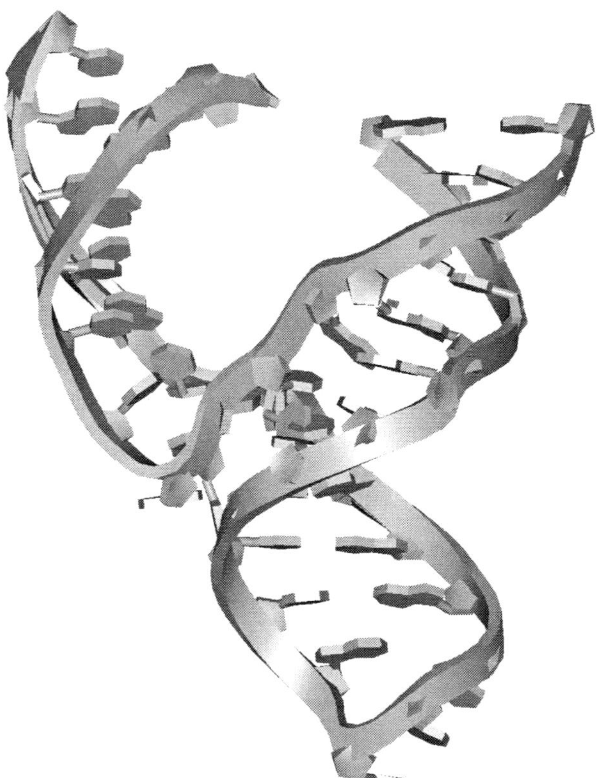

Fig. 6.16   The crystal structure of a hammerhead ribozyme.[42,43]

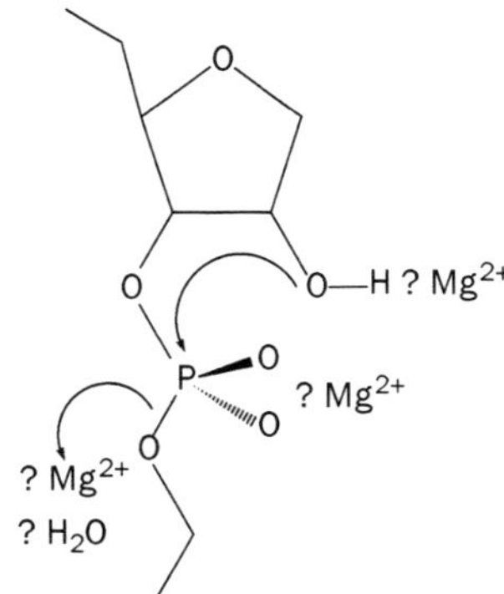

Fig. **6.17**   Schematic view of possible mechanisms of backbone cleavage by a hammerhead ribozyme. Sites of possible magnesium ion and water molecule involvement are indicated.

tandem A•G base pairs. This performs the key structural role of maintaining the correct geometric relationships between cleaved and catalytic strands.

The overall mechanism of hammerhead ribozyme catalysis involves nucleophilic attack by a 2′-hydroxyl group on the phosphorus atom of the adjacent backbone (Fig. 6.17). However elucidation of the details of the mechanism has not been straightforward, although it was recognized early on that metal ions play an important role, in view of the requirement for magnesium ions. The location of magnesium ions close to the A•G base pairs suggests that they provide a directly stabilizing influence on the precise conformation needed for catalysis. Several of the models for catalysis that have been proposed,[44] suggest that the ions are more directly involved. Magnesium ion(s) have been located at various positions within the active site, suggestive of a role in aiding the movement of electrons, perhaps by stabilization of leaving groups. Several crystal structures, together with evidence from other biophysical methods, have provided data relevant to the mechanistic questions.[45–47] Terbium ions inhibit catalytic cleavage, and the structure of a terbium complex[46] shows this ion to be located close to residues G5 and A6, and not within the active site. The structure of an intermediate has been captured,[47] which suggests that there is large-scale conformational change during the reaction, that brings the cleavage site and G5/A6 closer together than found in the ground-state structures.[42,43]

6.4.2 **Complex ribozymes**

The crystal structure has been determined of a Group I ribozyme domain, the P4-P6 domain of the *Tetrahymena* intron.[48] This structure (Fig. 6.18), contains 160 nucleotides, and is arranged as two roughly colinear helical regions. Within these are a number of RNA motifs, notably an adenosine-rich bulge, a three-way junction and a GAAA tetraloop. The close packing of the helices in this and other large RNA structures is due to the highly favourable interactions between unpaired, flipped-out adenosines attached to one helix and the minor groove of the adjacent one.[49] This provides an explanation for the high abundance of conserved unpaired adenosine residues in a number of large RNA molecules.

The P4-P6 domain does not contain the complete catalytic active site, so the structure does not give direct information on the mechanism of catalysis. The structure of the 247-nucleotide containing both domains contributing to catalysis has been described, at 5 Å resolution.[50] Even at this low resolution, the structure of the P4-P6 domain within it is unchanged from that in the P4-P6 structure itself. The 2nd domain in the larger structure, the P3-P9 domain, is on the exterior of the P4-P6 domain, making a large number of close contacts with it, involving base triplets and quartets. An important finding from this structure analysis is the high degree of accord between the experimentally

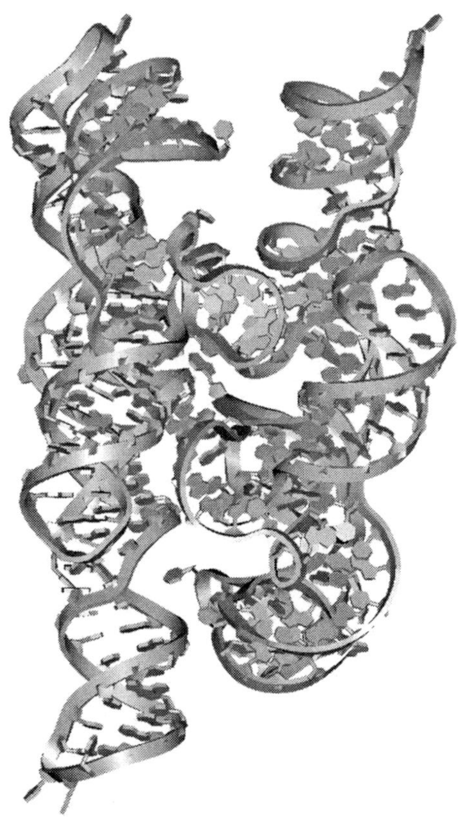

Fig. 6.18 The structure of the P4-P6 domain of the *Tetrahymena* intron.[48]

determined fold of the RNA with a three-dimensional model derived from biochemical and sequence comparison data.[51]

The crystal structures of the hepatitis delta virus and a hairpin ribozyme both show features relevant to their RNA-cleavage activities. The hepatitis delta virus[52] has a very complex arrangement of five helical moieties forming a double pseudo-knot arrangement (Fig. 6.19). (The pseudo-knot is another recurrent motif in RNA structure that involves the hydrogen-bonded interaction of a single-stranded region with a loop (Fig. 6.20).) This arrangement is fundamentally distinct from that in other ribozymes, as is the active site, which is buried within the five-helical core (Fig. 6.21). Another complex catalytic site

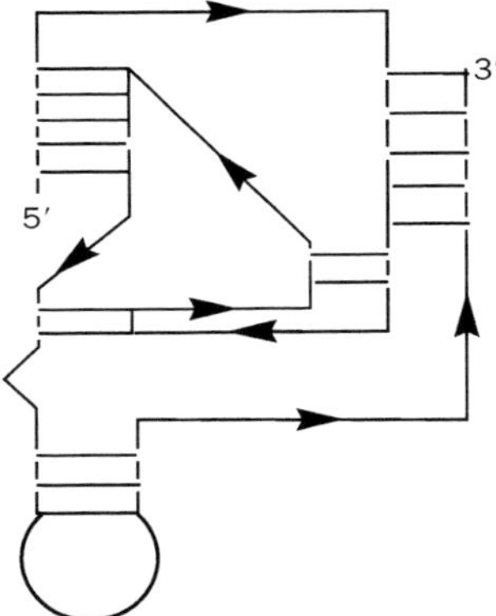

Fig. 6.19 Topology of the hepatitis delta virus ribozyme.[52]

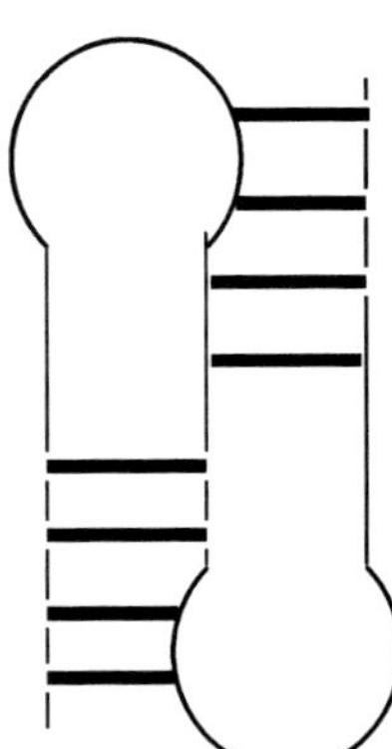

Fig. 6.20 Topology of a pseudo-knot.

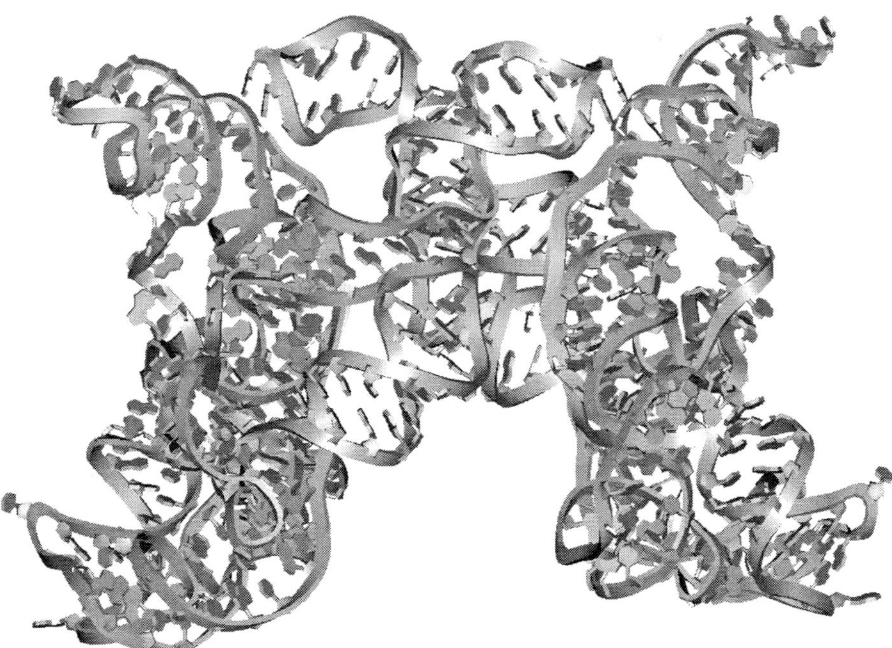

Fig. 6.21 Crystal structure of the hepatitis delta virus ribozyme.[52]

is seen in the structure of the hairpin ribozyme;[53] both have features resembling enzymes that also use activated phosphates, such as ribonuclease S.[54] The hairpin ribozyme has a four-way junction forming the central feature of the structure. The structures of several of the individual stems forming the junction have been previously individually studied; it is a salutary lesson in the complexities of RNA structure that these have undergone large conformational changes when a part of the complete hairpin ribozyme.

6.5 The ribosome, a ribozyme machine

The ribosome is responsible for protein synthesis in all prokaryotic and eukaryotic cells. It consists of two subunits, each a complex of proteins and ribosomal RNA. The complete 70S prokaryotic ribosome has a total molecular weight of *c.* 2.5 million Daltons. In prokaryotics the 30S subunit contains about 20 proteins and a single RNA molecule of around 1500 nucleotides in length. The larger 50S subunit contains half as many more proteins and a large RNA of about 3000 nucleotides, together with the small (120 nucleotide) 5S RNA. The primary function of the small subunit is to control tRNA interactions with messenger RNAs. The large subunit controls peptide transfer and undertakes the catalytic function of peptide bond formation.

Structural studies on bacterial ribosomes have been underway for almost 40 years, with the ultimate goal of achieving atomic resolution in order to understand the mechanics of ribosome function—for most of that time appearing to be an impossibly difficult task. A number of individual ribosomal proteins have been studied, but their three-dimensional arrangement within the ribosome, and their functional roles, have not been apparent. Electron microscopy has enabled the outlines of ribosome structure to be defined, which have been of considerable help to the recent single-crystal analyses. Crystals of diffraction quality were first obtained over 20 years ago, but only recently have the various necessary technologies become available for tackling such a massive and complex crystallographic task. It was necessary to collect huge data sets from the very large unit cells involved, with crystals that diffracted only weakly and decayed rapidly in the X-ray beam. This has been achieved using synchrotron radiation together with low-temperature methods to stabilize the crystals from excessive decay.

Determining the phases of these large structures has also been a formidable challenge. Rather than using single heavy atoms, the use of clusters of heavy metals has been successful, especially for the initial structures, at less than atomic resolution. The progress of ribosome structure determination has been rapid (Table 6.6). The first structure, at 9 Å resolution, was reported in 1998,[55] culminating most recently in electron-density maps of the small[60] and large[61] subunits at 3.0 and 2.4 Å resolution respectively, at which point individual side chains become apparent and can be refined. These structures are the largest yet determined by X-ray crystallography, and dissection of the electron-density maps has only been possible with the aid of the large body of biochemical and other data accumulated on the ribosome and its subunits. It is remarkable that the secondary structures for the RNAs in these complexes are close to those predicted by analyses of RNA across species. This phylogenetic analysis is based on the reasonable assumption that RNA structure is largely conserved throughout evolution.

A number of ribosome coordinate data sets are now available from the Data Banks. Although only backbone C_α carbon atoms were deposited for the protein components in the earlier lower-resolution structures, side chain atoms have been identified in the more

Table 6.6 Progress in ribosome crystallographic structure determination

Subunit	Organism	Resolution (Å)	Year	Reference
50S	*H. marismortui*	9	1998	55
50S	*H. marismortui*	5.0	1999	56
30S	*T. thermophilus*	5.5	1999	57
70S	*T. thermophilus*	7.8	1999	58
30S	*T. thermophilus*	4.5	1999	59
30S	*T. thermophilus*	3.0	2000	60
30S	*T. thermophilus*	3.3	2000	66
50S	*H. marismortui*	2.4	2000	61
70S	*T. thermophilus*	5.5	2001	62

Table 6.7 Selected ribosome subunit crystal structures

Subunit and resolution	NDB number
30S, 3.0 Å	RR0016
50S, 2.4 Å	RR0031
30S with mRNA +tRNA, 3.3 Å	RR0029, RR0030

recent analyses. The size and complexity of ribosome structures is such that there is some advantage when viewing these very large assemblies, in restricting initial viewing to the nucleic acid components alone (Table 6.7). The reader is referred to this book's web site (http://www.oup.com/na-structure) for detailed views of the subunits and the complete ribosome. The necessarily brief survey below of the structures cannot do justice to their complexity, and the reader is strongly encouraged to read the primary literature on them. The structure of the complete 70S ribosome,[62] although still at relatively low resolution, reveals much about the interactions between the subunits that are an essential element of the protein synthesis cycle. It is notable that even though the resolution as yet precludes any detailed study of the interactions involved, the three bound tRNA molecules are all extensively contacted by ribosomal RNA in addition to the necessary established functional interactions such as codon–anticodon recognition.

6.5.1 The structure of the 30S subunit

The 3.0 Å crystal structure[60] has located the complete 16S ribosomal RNA of 1511 nucleotides together with the ordered regions of 20 ribosomal proteins, altogether organized into four well-defined domains. The implication is that there is considerable flexibility between them, which is needed in order to ensure the movement of messenger and transfer RNAs. The overall shape of the 30S particle is dominated by the structure of the folded RNA (Fig. 6.22), and the proteins serve merely to fill up the gaps and hold the whole assembly together. The secondary structure of the RNA shows over 50 helical regions. The numerous loops are mostly small and do not disrupt the runs of helix in which they are embedded. There are extensive interactions between helices, mostly involving co-axial stacking via the minor grooves. In one type of helix–helix interaction two

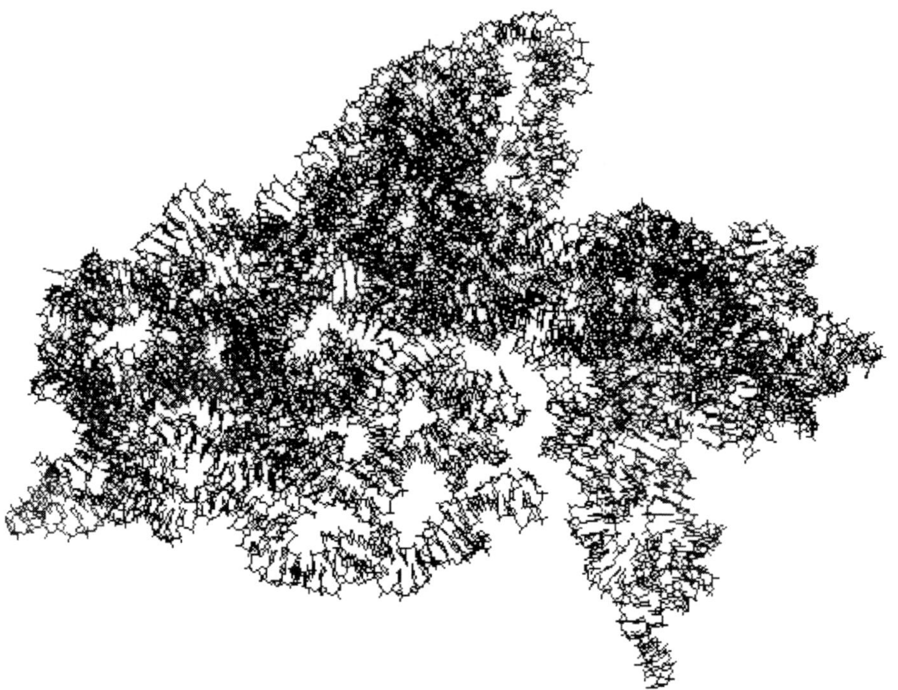

Fig. 6.22 Crystal structure of the RNA component in the 30S ribosome subunit.[60]

minor grooves abut each other, with consequent distortions from A-type geometry. These distortions, which tend to involve runs of adenines, are facilitated by both extra-helical bulges and non-canonical base pairs, as have been observed in simple RNA structures. Less commonly, perpendicular packing of one helix against another (also via the minor groove) is mediated by an unpaired purine base. This mode is of special importance since it involves the functionally significant helices in the 30S subunit. The motifs of RNA tertiary structure such as non-Watson–Crick base pairs, base triplets and tetraloops, all contribute to the overall structure.

The majority of the 20 ribosomal proteins in the structure each consist of a globular region and a long flexible arm. The latter have been too flexible to be observed in structural studies on the individual proteins, but have been located in the 30S subunit, where they play important roles in helping to stabilize the RNA folding, by essentially filling in the numerous spaces in the RNA folds.

The structure identifies the three sites where tRNA molecules bind and function, and where the essential proof-reading checks for fidelity of code-reading and translation occur. These sites are:

(1) the P (peptidylation) site, when a tRNA anticodon base pairs with the appropriate codon in mRNA, and where the peptide chain is covalently linked to a tRNA;

(2) the A (acceptor) site, when peptide bonds are eventually formed (the actual peptidyl transferase steps occur in the 50S subunit);

(3) the E (exit) site, for tRNAs to be released from the subunit as part of the protein synthesis cycle.

tRNA itself is not present in this crystal structure, but the RNA from a symmetry-related 30S subunit effectively serves to mimic it as the anticodon stem-loop. Its interactions with the 30S RNA are also mediated via (i) minor-groove surfaces, helped by some contacts with ribosomal proteins, and (ii) via backbone contacts. Interestingly, the exit site of the tRNA is almost exclusively protein-associated, whereas the other functional sites are composed of RNA and not protein. It is thus the ribosomal RNA that mediates the functions of the 30S subunit, and not the ribosomal proteins.

The determination[64] of the crystal structure at 3.3 Å of the 30S subunit complete with the anticodon stem-loop of an tRNA and a short RNA acting as an mRNA, has enabled the details of codon–anticodon recognition and checking to be visualized. A key role is played by two adjacent adenines from the 16S RNA, which are in contact with the minor groove of the codon–anticodon helical stem, so as to verify the correctness of the first two Watson–Crick base pairs of the triplet, and selection of the correct codon. The third codon position (the 'wobble' position), is not as involved in interactions with the 16S RNA and a tRNA anticodon is thus able to cope with hydrogen bonding to a range of 3rd-position codon bases. The codon–anticodon stringency is disrupted by the antibiotic paromomycin, which has also been co-crystallized with the 30S subunit. Several other antibiotics that are known to inhibit protein synthesis have been examined bound to the 30S subunit, following co-crystallization experiments.[63,65] For example, streptomycin has been found to bind tightly to four distinct sites on the 16S RNA, mostly via phosphate groups. This stabilizes the A site in an open conformation, thereby enabling non-cognate tRNAs to bind and reducing proof-reading capability. These and other studies will surely enable new and more effective antibiotics to be rationally designed, and current clinical problems of antibiotic resistance to be addressed at a molecular level. A second crystal structure,[66] also at 3.3 Å resolution, has shown that the decoding region of the structure, is highly conserved. Furthermore the A and P site tRNAs and the codon part of the bound mRNA do not interact with any of the proteins, again emphasizing the dominant role of RNA in ribosome structure and function.

6.5.2 **The structure of the 50S subunit**

In the 2.4 Å crystal structure[61] 2711 out of the total of 2923 ribosomal RNA nucleotides have been observed, together with 27 ribosomal proteins, and 122 nucleotides of 5S RNA, with the subunit being about 250 Å in each dimension. Again, the function of the proteins appears to be mainly to act as a cement, helping to maintain the integrity of the folded RNA. Although the RNA itself can be arranged into six domains on the basis of its secondary structure, overall the 50S subunit is remarkably globular (Fig. 6.23), reflecting its greater conformational rigidity compared to the 30S subunit, consist with its functional need to be more flexible.

The principal function of the 50S subunit is peptide bond formation. The active site where this occurs is within domain 5. Its precise location was identified by the analysis of crystals containing the antibiotic puromycin, an inhibitor of peptide synthesis. Remarkably, there is no protein structure within 18 Å of this molecule, bound in its active site, and accordingly it has been proposed that peptidyl transferase catalytic activity is entirely carried out by RNA, analogous to the function of RNAs as ribozymes. The reaction is an acid-base hydrogen transfer, with adenine being suggested as the key residue, accepting a hydrogen atom from an aminoacylated tRNA so that it can bond to the carbonyl group of an

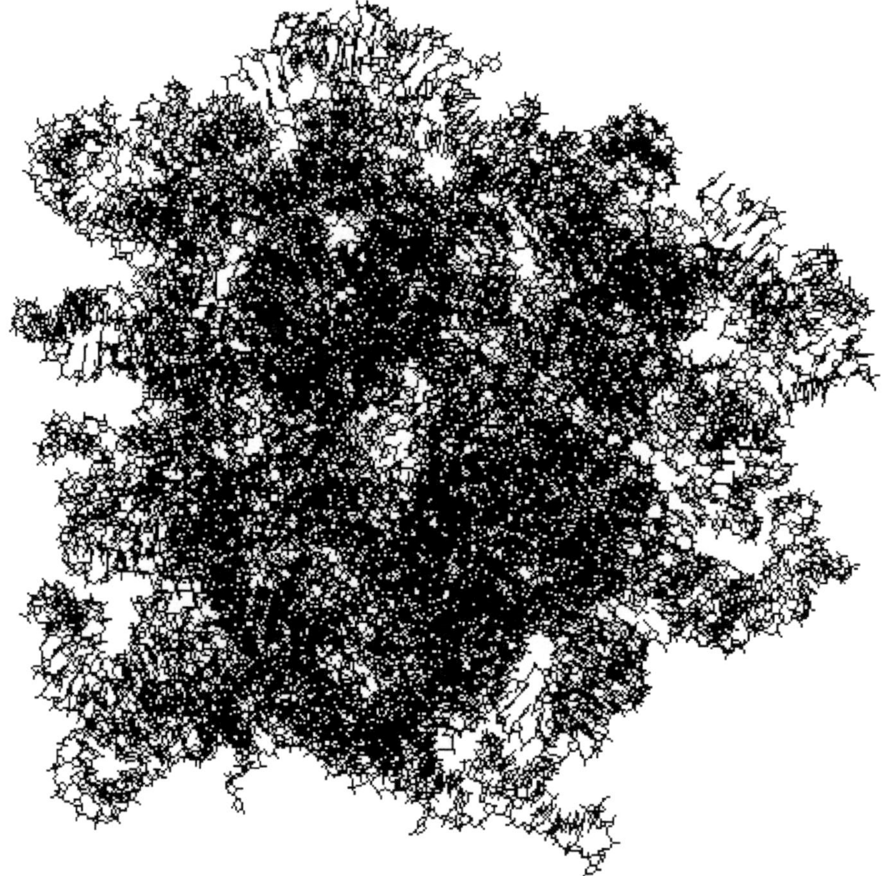

Fig. 6.23 Crystal structure of the RNA component in the 50S ribosome subunit.[61]

esterified peptidyl-tRNA carrying the peptide chain to which the new amino acid is to be attached. All nucleotides involved in the active site are highly conserved in nature, strongly suggesting that RNA catalysis is a fundamental mechanism acting throughout evolution.

6.6 **RNA motifs**

We have seen in this chapter that the variability and complexity of RNA structures is due to their ability to fully exploit their hydrogen bonding and base stacking potential. In hydrogen bonding terms Watson–Crick base pairing remains the bedrock of helical stems, essential components of all RNAs. In addition, RNAs exploit the full gamut of potential base pairing presented by purines and pyrimidines, in ways that very much extend the mismatches seen with DNA, as discussed in Chapter 4. These pairings have been systematized and classified into 12 families in a rational way[67] that emphasizes the three edges of a base—Watson–Crick, Hoogsteen, and sugar/base. There is even greater diversity of possible base triplets and quartets, that remain to be fully catalogued, once the ribosome structures, which are the richest mines of information on RNA tertiary folds, progress to increasingly higher resolution. These interactions in turn are the basis of the various established RNA motifs, the adenosine platform, various tetraloops and the pseudo-knot. It is already apparent that the sheer density of nucleotides in the ribosome results in a very large number of loop insertions into helices, primarily via their minor grooves, and a number of novel structural motifs are involved. One, the A-minor motif, appears to be specially prevalent.[68] This involves adenines interacting in the minor groove, of which four distinct ways of doing so have been identified. The A-minor motif is an important component of the manner in which ribosomal RNA preserves its functionally significant architecture—the adenines involved are often highly conserved, and so the ribosomal RNA framework is likely to be largely invariant throughout the living world. The motif also appears in several ribozyme structures (see also Section 6.4.2).

References

1   Kundrot, C. (1997). *Methods in Enzymology*, **276**, 143.

2   Anderson, A. C., Scaringe, S. A., Earp, B. E., and Frederick, C. A. (1996). *RNA*, **2**, 110.

3   Price, S. R., Ito, N., Oubridge, C., Avis, J. M., and Nagai, K. (1995). *Journal of Molecular Biology*, **249**, 398.

4   Doudna, J. A., Grosshans, C., Gooding, A., and Kundrot, C. E. (1993). *Proceedings of the National Academy of Sciences USA*, **90**, 7829.

5   Cate, J. H. and Doudna, J. A. (2000). *Methods in Enzymology*, **317**, 169.

6   Dock-Breigon, A. C., Moras, D., and Giegé, R. (1999). In *Crystallisation of Nucleic Acids and Proteins* (eds. Ducruix, A. and Giegé, R.,), Oxford University Press, Oxford.

7   Ferre-D'Amare, A.R., Zhou, K., and Doudna, J. A. (1998). *Journal of Molecular Biology*, **279**, 621.

8   Wedekind, J. E. and McKay, D. B. (2000). *Methods in Enzymology*, **317**, 149.

9   Auffinger, P. and Westhof, E. (1997). *Journal of Molecular Biology*, **274**, 54.

10   Arnott, S., Hukins, D. W. L., Dover, S. D., Fuller, W., and Hodgson, A. R. (1973). *Journal of Molecular Biology*, **81**, 107.

11   Arnott, S. (1999) In *Oxford Handbook of Nucleic Acid Structure* (ed. Neidle, S.), p. 1, Oxford University Press, Oxford.

12   Seeman, N. C., Rosenberg, J. M., Suddath, F. L., Kim, J. J. P., and Rich, A. (1976). *Journal of Molecular Biology*, **104**, 109.

13   Rosenberg, J. M., Seeman, N. C., Day, R. O., and Rich, A. (1976). *Journal of Molecular Biology*, **104**, 145.

14   Portmann, S., Usman, N., and Egli, M. (1995). *Biochemistry*, **34**, 7569.

15   Dock-Bregeon, A. C., Chevrier, B., Podjarny, A., Johnson, J., de Bear, J. S., Gough, G. R., *et al.* (1989). *Journal of Molecular Biology*, **209**, 459.

16   Schindelin, H., Zhang, M., Baid, R., Fürste, J.-P., Erdmann, V. A., and Heinemann, U. (1995). *Journal of Molecular Biology*, **249**, 595.

17   Klosterman, P. S., Shah, S. A., and Steitz, T. A. (1999). *Biochemistry*, **38**, 14784.

18   Egli, M., Portmann, S., and Usman, N. (1996). *Biochemistry*, **35**, 8489.

19   Auffinger, P. and Westhof, E. (2000). *Journal of Molecular Biology*, **300**, 1113.

20   Cheatham, T. E. and Kollman, P. A. (1997). *Journal of the American Chemical Society*, **119**, 4805.

21   Cruse, W. B. T., Saludjian, P., Biala, E., Strazewski, P., Prangé, T., and Kennard, O. (1994). *Proceedings of the National Academy of Sciences USA*, **91**, 4160.

22   Holbrook, S. R., Cheong, C., Tinoco, I., and Kim, S.-H. (1991). *Nature*, **353**, 579.

23   Baeyens, K. J., de Bondt, H. L., Pardi, A., and Holbrook, S. R. (1996). *Proceedings of the National Academy of Sciences USA*, **93**, 12851.

24   Heus, H. A. and Pardi, A. (1991). *Science*, **253**, 191.

25   Varani, G. and McClain, W. H. (2000). *EMBO Reports*, **1**, 18.

26   Ennifar, E., Yusupov, M., Walter, P., Marquet, R., Ehresmann, B., Ehresmann, C., and Dumas, P. (1999). *Structure*, **7**, 1439.

27   Wild, K., Weichenrieder, O., Leonard, G. A., and Cusack, S. (1999). *Structure*, **7**, 1345.

28   Peterson, R. D. and Feigon, J. (1996). *Journal of Molecular Biology*, **264**, 863.

29   Ippolito, J. A. and Steitz, T. A. (2000). *Journal of Molecular Biology*, **295**, 711.

30   Allain, F. H.-T. and Varani, G. (1995). *Journal of Molecular Biology*, **250**, 333.

31   Ennifar, E., Nikulin, A., Tishchenko, S., Serganov, A., Nevskaya, N., Garber, M., *et al.* (2000). *Journal of Molecular Biology*, **304**, 35.

32   Miller, J. L. and Kollman, P. A. (1997). *Journal of Molecular Biology*, **270**, 436.

33   Rife, J. P., Stallings, S. C., Correll, C. C., Dallas, A., Steitz, T. A., and Moore, P. B. (1999). *Biophysical Journal*, **76**, 65.

34   Robertus, J. D., Ladner, J. E., Finch, J. T., Rhodes, D., Brown, R. S., Clark, B. F. C., and Klug, A. (1974). *Nature*, **250**, 546.

35   Kim, S.-H., Suddath, F. L., Quigley, G. J., McPherson, A., Sussman, J. L., Wang, A. H.-J., *et al.* (1974). *Science*, **185**, 435.

36   Westhof, E. and Sundaralingam, M. (1986). *Biochemistry*, **25**, 4868.

37   Westhof, E., Dumas, P., and Moras, D. (1988). *Acta Crystallographica*, **A44**, 112.

38   Shi, H. and Moore, P. B. (2000). *RNA*, **6**, 1091.

39   Jovine, L., Djordjevic, S., and Rhodes, D. (2000). *Journal of Molecular Biology*, **301**, 401.

40   Cech, T. R., Zaug, A. J., and Grabowski, P. J. (1981). *Cell*, **27**, 487.

41   Guerrier-Takada, C., Gardiner, K., Marsh, T., Pace, N., and Altman, S. (1983). *Cell*, **35**, 849.

42   Pley, H. W., Flaherty, K. M., and McKay, D. B. (1994). *Nature*, **372**, 68.

43   Scott, W. G., Finch, J. T., and Klug, A. (1995). *Cell*, **81**, 991.

44   Lilley, D. M. J. (1999). *Current Opinion in Structural Biology*, **9**, 330.

45   Murray, J. B., Terwey, D. P., Maloney, L., Karpeisky, A., Usman, N., Beigelman, L., and Scott, W. G. (1998). *Cell*, **92**, 665.

46   Feig, A. L., Scott, W. G., and Uhlenbeck, O. C. (1998). *Science*, **279**, 81.

47   Murray, J. B., Szöke, H., Szöke, A., and Scott, W. G. (2000). *Molecular Cell*, **5**, 279.

48   Cate, J. H., Gooding, A. R., Podell, E., Zhou, K., Golden, B. L., Kundrot, C. E., *et al.* (1996). *Science*, **273**, 1678.

49   Doherty, E. A., Batey, R. T., Masquida, B., and Doudna, J. A. (2001). *Nature Structural Biology*, **8**, 339.

50   Golden, B. L., Gooding, A. R., Podell, E. R., and Cech, T. R. (1998). *Science*, **282**, 259.

51   Lehnert, V., Jaeger, L., Michel, F., and Westhof, E. (1996). *Chemistry and Biology*, **3**, 993.

52   Ferré-D'Amaré, A. R., Zhou, K., and Doudna, J. A. (1998). *Nature*, **395**, 567.

53   Rupert, P. B. and Ferré-D'Amaré, A. R. (2001). *Nature*, **410**, 780.

54   Strobel, S. A. and Ryder, S. P. (2001). *Nature*, **410**, 761.

55   Ban, N., Freeborn, B., Nissen, P., Penzyek, P., Grassucci, R. A., Sweet, R. M., *et al.* (1998). *Cell*, **93**, 1105.

56   Ban, N., Nissen, P., Hansen, J., Capel, M., Moore, P. B., and Steitz, T. A. (1999). *Nature*, **400**, 841.

57   Clemons, W. M., May, J. L. C., Wimberley, B. T., McCutcheon, J. P., Capel, M. S., and Ramakrishnan, V. (1999). *Nature*, **400**, 833.

58   Cate, J. H., Yusupov, M. M., Yusupova, G. Z., Earnest, T. N., and Noller, H. F. (1999). *Science*, **285**, 2095.

59   Tocilj, A., Schlunzen, F., Janell, D., Gluhmann, M., Hansen, H. A. S., Harmo, J., *et al.* (1999). *Proceedings of the National Academy of Sciences USA*, **96**, 14252.

60   Wimberley, B. T., Brodersen, D. E., Clemons, W. M., Morgan-Warren, R. J., Carter, A. P., Vonrein, C., *et al.* (2000). *Nature*, **407**, 327.

61   Ban, N., Nissen, P., Hansen, J., Moore, P. B., and Steitz, T. A. (2000). *Science*, **289**, 905.

62   Yusupov, M. M., Yusupova, G. Z., Baucom, A., Lieberman, K., Earnest, T. N. Cate, J. H. D., and Noller, H. F. (2001). *Science*, **292**, 883.

63   Carter, A. P., Clemons, W. M., Brodersen, D. E., Morgan-Warren, R. J., Wimberley, B. T., and Ramakrishnan, V. (2000). *Nature*, **407**, 340.

64   Ogle, J. M., Broderson, D. E., Clemons, W. M., Tarry, M. J., Carter, A. P., and Ramakrishnan, V. (2001). *Science*, **292**, 897.

65   Brodersen, D. E., Clemons, W. M., Carter, A. P., Morgan-Warren, R. J., Wimberley, B. T., and Ramakrishnan, V. (2000). *Cell*, **103**, 1143.

66   Schlunzen, F., Tocilj, A., Zarivach, R., Harms, J., Gluhmann, M., Janell, D., *et al.* (2000). *Cell*, **102**, 615.

67   Leontis, N. B., and Westhof, E. (2001). *RNA*, **7**, 499.

68   Nissen, P., Ippolito, J. A., Ban, N., Moore, P. B., and Steitz, T. A. (2001). *Proceedings of the National Academy of Sciences USA*, **98**, 4899.

Further reading

RNA structures

Batey, R. T., Rambo, R. P., and Doudna, J. A. (1999). *Angewante Chemie, International Edition*, **38**, 2326.

Shen, L. X., Cai, Z., and Tinoco, I. (1995). *FASEB Journal*, **9**, 1023.

Westhof, E. (1999). In *Oxford Handbook of Nucleic Acid Structure* (ed. Neidle, S.), p. 533, Oxford University Press, Oxford.

tRNA

Arnez, J. G. and Moras, D. (1999). In *Oxford Handbook of Nucleic Acid Structure* (ed. Neidle, S.), p. 603, Oxford University Press, Oxford.

Kim, S.-H. (1981). In *Topics in Nucleic Acid Structure*, (ed. Neidle, S.), p. 83, Macmillan Press, London.

Ribozymes

Butcher, S. E. (2001). *Current Opinion in Structural Biology*, **11**, 315.

Doherty, E. A. and Doudna, J. A. (2001). *Annual Reviews in Biophysics*, **30**, 457.

Ferré-D'Amaré, A. R. and Doudna, J. A. (1999). *Annual Reviews in Biophysics*, **28**, 57.

Takagi, Y., Warashina, M., Stec, W. J., Yoshinari, K., and Taira, K. (2001). *Nucleic Acids Research*, **29**, 1815.

Ribosome structure

Cech, T. R. (2000). *Science*, **289**, 878.

Maguire, B. A. and Zimmerman, R. A. (2001). *Cell*, **104**, 813.

Moore, P. B. (2001). *Biochemistry*, **40**, 3243.

Web sites of interest

http://www-lbit.iro.umontreal.ca/mcsym/ This site gives access to the Mc-Sym and Mc-Annotate programs for building and analysing RNA 3D structures (see Gendron, P., Lemieux, S., and Major, F. (2001). *Journal of Molecular Biology*, **308**, 919).

http://www.rnabase.org This site provides data on all publicly available three-dimensional RNA structures, taking coordinates that are deposited in the NDB. RNAbase also provides conformational maps and lists of parameter outliers.

http://chem.bgsu.edu/RNA This site provides detail on non-canonical base pairs, which are classified on the basis of the scheme in Leontis, N. B. and Westhof, E. (1998). *Quarterly Reviews of Biophysics*, **31**, 399.

7

Principles of protein–DNA recognition

7.1 Introduction

The sequence information within DNA exists to be replicated, read and translated into gene products, depending on the nature of the cell type, its progression through the cell cycle, and its response to external stimuli and factors. The processes of packaging DNA in the cell, regulating the expression of individual genes, and then reading/translating the encoded sequences, requires the interplay of large numbers of non-specific and specific proteins. It is thus unsurprising that the study of DNA–protein interactions has developed into a major area in its own right, of which structural studies are one important part. This chapter concentrates on the underlying principles of recognition as found by structural methods (principally X-ray crystallography), and how DNA structure itself copes with a wide range of protein and functional requirements. The reader is directed to the extensive primary literature for detailed descriptions of individual structures.

Structural information is now available on over 400 distinct DNA–protein complexes, from a wide range of eukaryotic and prokaryotic sources. Studies of the proteins themselves rarely provide sufficient insight into the processes of recognition, and are not discussed here. This relatively small number of known structures pales into insignificance compared to the total encoded by individual genomes. The determination of the human genome sequence in 2001 has enabled reliable estimates to be made for the numbers of genes with particular functions. Of the *c.* 30000 in total, 13.5 per cent (2308) are proposed to be involved in nucleic acid binding, of which 6 per cent (1850) are estimated to be transcription factors.[1] We shall undoubtedly see over the next decade a major effort in large-scale high-throughput structural studies (structural genomics), that will greatly reduce this disparity between numbers of known and unknown structures. However, major increases in our understanding of DNA–protein recognition and processing will not be obtained solely from complexes with single proteins, but from studies of functionally relevant multi-protein complexes, as we have seen in the case of the ribosome.

DNA-binding proteins can be conveniently categorized in both functional and structural terms into a small number of families. The principal functional groups are:

1. Regulatory proteins. These mostly bind to highly specific sequences of duplex DNA, in order to control the transcription of a particular gene, or bind to particular signal sequences such as 5′-TATA in order to more generally initiate transcription.
2. DNA cleavage proteins (nucleases). Some such as DNase I have relatively little sequence specificity (although they may have some DNA structural selectivity). Others, the restriction enzymes, are highly specific for particular sequences.
3. Repair proteins that respond to various types of damage to DNA by recognizing the lesion itself, then excising the damaged DNA and/or joining together breaks in damaged DNA.
4. Proteins that resolve topological problems in DNA by unravelling or unwinding DNA prior to replication. The DNA topoisomerases, of importance as cancer therapeutic targets, have been extensively studied.
5. Structural proteins that maintain the integrity of folded or packaged DNA, for example, histones in chromatin.
6. Processing proteins, typified by DNA and RNA polymerases, that use DNA as a template for further nucleic acid synthesis. Sequence specificity is absolutely not required for DNA recognition, rather a need to recognize a particular type of DNA duplex.

This diversity of functions shown by DNA-binding proteins is in striking contrast to the relatively few ways in which DNA structure is 'read', even though there is a very wide variety in the structure of the proteins themselves. Direct reading of a DNA sequence generally occurs via the hydrogen-bonding edges of the bases,[2] as described in Chapter 5 for small molecules; therefore, features are required of a protein that enable the bases to be accessed through either major or minor grooves. Figure 7.1 shows that the B-DNA major groove is the richer of the two grooves of duplex DNA, both in information content per se, and in its ability to facilitate discrimination between different DNA sequences, which is essential if the appropriate genes are to be transcribed. Thus, the major groove is generally the site of direct information readout. Nonetheless, the minor groove is an important target for some regulatory and structural proteins, especially those that are able to deform DNA so that the minor groove becomes greatly expanded.

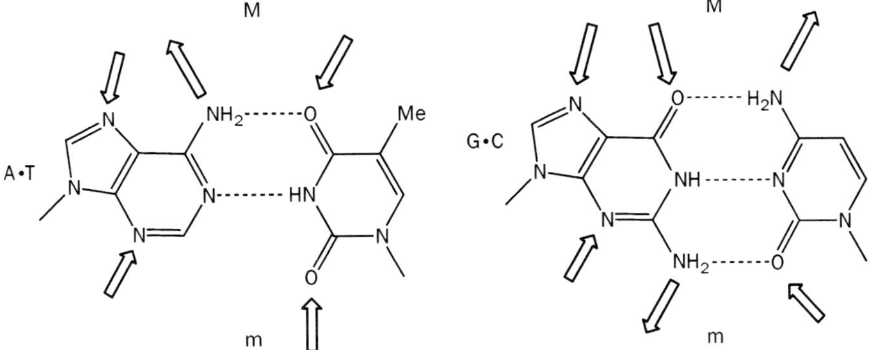

Fig. 7.1 The pattern of hydrogen-bond donors and acceptors in Watson–Crick base pairs. The direction of the arrows indicates the direction of hydrogen-bonding donation.

Indirect readout can occur via the sugar–phosphate backbone or solely the phosphate groups. Especially when this is sequence-specific, it is the cumulative consequence of a number of individually non-specific, non-bonded interactions as compared to the clearly defined hydrogen bonding interactions involved in direct recognition. The anionic nature of phosphate groups makes them frequent targets for basic side chains, with the resulting electrostatic interactions being significant contributors to overall protein–DNA binding.

There are a limited number of ways by which proteins recognize duplex DNA.[3,4] To some extent this is unsurprising given that the majority of protein–DNA complexes retain DNA in overall B-type conformations. The principal recognition motifs are:

- Helix-turn-helix
- Zinc-finger and loop
- Zipper motifs
- β-sheets
- β-hairpins

These are described below.

7.2 Direct protein–DNA contacts

These interactions involve hydrogen bonds between amino acid side chains and the edges of the base pairs, involving their pattern of donors and acceptors molecules. The same principles of hydrogen bonding apply as for small molecules binding to DNA (see Chapter 5), although the common contact areas involved in protein–DNA interfaces are invariably much larger.[5] The protein backbone itself does sometimes make contact with bases or phosphates, but these are generally not determinants of specificity, rather they serve to enhance binding affinity. The majority of interactions involve O6 and/or N7 atoms of guanine bases forming hydrogen bonds with the charged ends of long flexible side chains from the basic residues arginine or lysine, the amide residues glutamine and asparagine or the hydroxyl group of a serine (Table 7.1 and Figs 7.2–7.3). The recognition of one hydrogen-bonding site on a base, compared to two simultaneous sites on that base, actually involves only a small change in position of the amino acid side chain. In general, the latter will be energetically preferred and formed if possible. Adenine bases are recognized via their N6 and/or N7 atoms, although this occurs less frequently than guanine recognition. Active pyrimidine recognition is also less common than guanine recognition.

Table 7.1 Distribution of amino acid–base interactions in protein–DNA, adapted from,[6] using a data set of 129 structures. Only those amino acids that participate in a significant number of DNA base interactions are listed

	Guanine	Cytosine	Adenine	Thymine
Arginine	98	8	19	24
Lysine	30	6	4	9
Serine	12	2	1	3
Asparagine	7	10	18	7
Glutamine	6	2	16	2
Glutamate	1	10	1	0

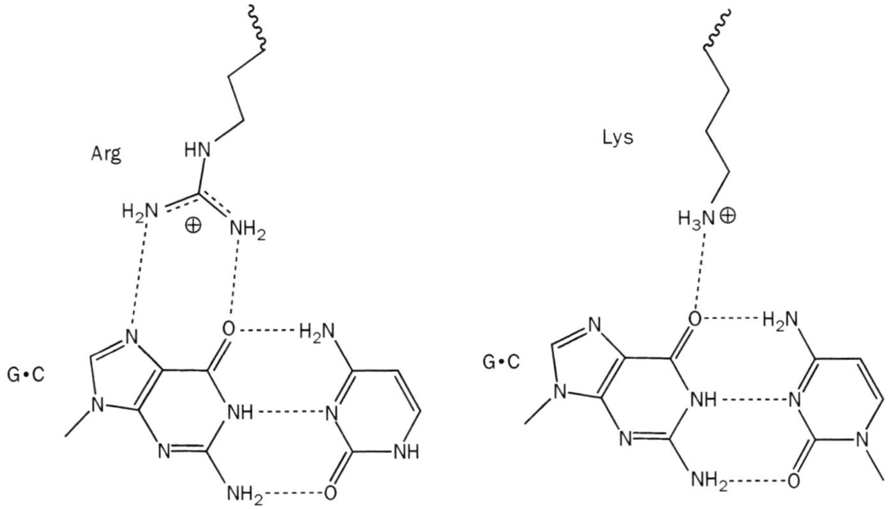

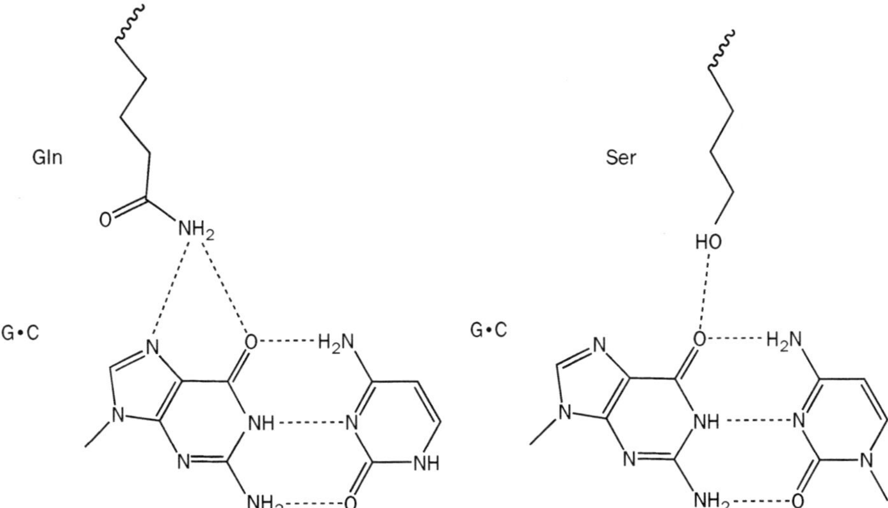

Fig. 7.2 Possible patterns of hydrogen-bonding recognition of G•C base pairs by arginine and lysine side chains.

Fig. 7.3 Possible patterns of hydrogen-bonding recognition of G•C base pairs by glutamine and serine side chains.

There is no general 1 : 1 amino acid : DNA base correspondence, and recognition can sometimes occur in a wide variety of ways in addition to the simple mono- or bidentate ones. Table 7.1 also shows that hydrogen bonding to both partners simultaneously in Watson–Crick base pairs is relatively uncommon, not least since binding to one usually results in sufficient discrimination for sequence selectivity to take place. Some arrangements for side chains bridging between two bases in a base pair are shown in Figs 7.4a,b. Hydrogen bonding can also occur to two consecutive bases on the same strand (Fig. 7.5). This arrangement introduces sequence selectivity at the dinucleotide level. For example, the TG sequence shown in Fig. 7.5, is recognized by the lysine side chain through both carbonyl groups of the

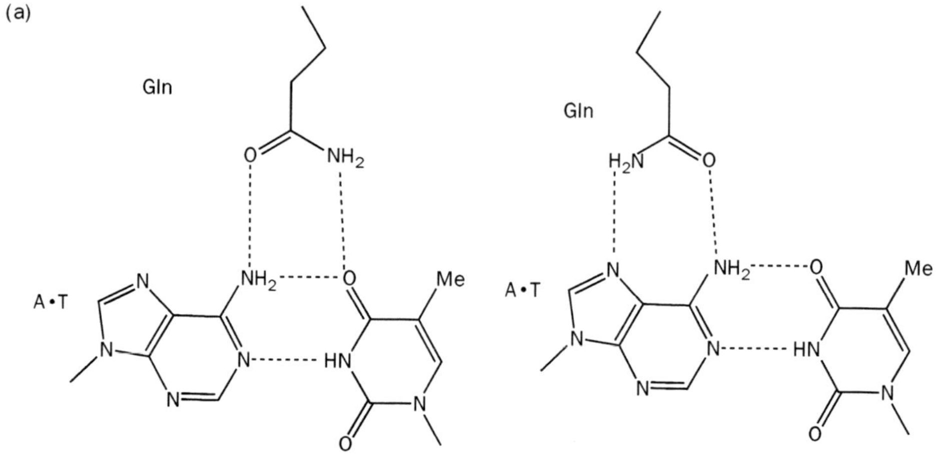

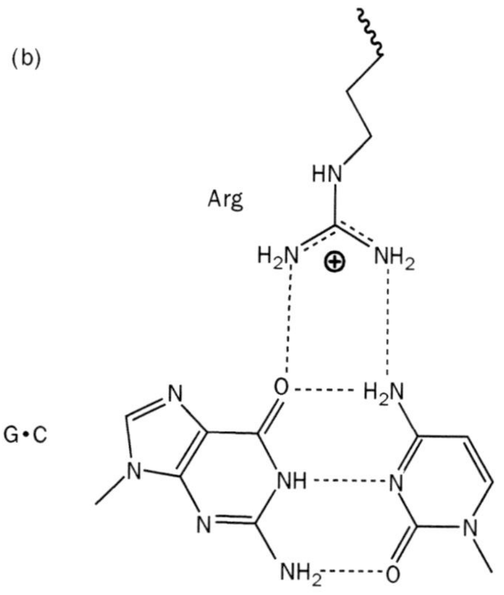

Fig. 7.4 (a) Two possible patterns of hydrogen-bonding recognition of A•T base pairs by glutamine side chains.; (b) Arginine recognition of both bases in a G•C base pair.

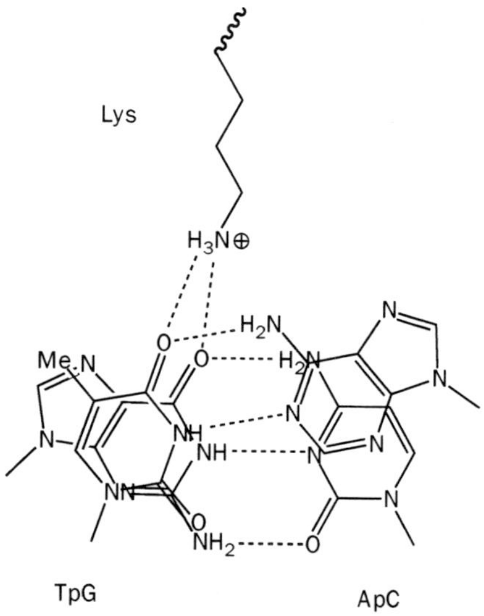

Fig. 7.5 Possible recognition by a lysine side-chain of the sequence TpG.

thymine and guanine bases. Altering this to TA or TC would not put two carbonyl groups in the same correct position for interaction with lysine. Similar arguments can be made for other dinucleotide steps involved in bidentate recognition with side chains.

Sometimes a water molecule (or molecules) participates in such a hydrogen-bonded bridge, or itself bridges between a base and a side chain. Such water involvement has been found in the *E. coli trp* repressor/operator complex,[7] with only two (relatively unimportant) direct (guanine · · · arginine) contacts, yet three that have water-mediated contacts. A further example of the important role that water molecules can play, is in the bacteriophage λ repressor–operator complex[8] where the hydroxyl group of serine-45 interacts directly with O6 and N7 of a guanine. The backbone carbonyl oxygen atom of this serine interacts via a water molecule with the N4 atom of the complementary cytosine base and through the same water molecule to O4 of the adjacent thymine.

It has not been possible to establish a pattern of preference between a particular base and an amino acid residue. This lack of a general hydrogen-bonding recognition pattern (although there are some involving very closely related proteins), reflects a number of factors (9–11):

(1) the effect of differences in DNA structure in various complexes;
(2) the different types of protein motif involved in DNA recognition;
(3) other recognition factors that can play significant roles in defining the recognition of particular sequences, notably the ability of thymine methyl groups to form close van der Waals attractive interactions with hydrophobic side chains;
(4) in general, any one base can be recognized by several amino acids.

An example of the role of thymine methyl groups is in the engrailed homeodomain structure, where there is a close contact between isoleucine-47 and one such methyl group from a thymine in the TAAT recognition site.[12] Important roles are played by numerous non-specific (at least in hydrogen-bonding terms) protein–backbone and other interactions in stabilizing a protein–DNA complex as a whole. Thus, the flexibility of amino acid side chains ensures that optimal interactions are made with all elements of a DNA structure. Taken as a whole, these enable the recognition helix to be oriented in the major groove in a manner that is distinctive for a particular protein. Such generalized hydrophobic and electrostatic interactions may also play a more active recognition role, by means of indirect readout, sensing the DNA structure and flexibility specified by a particular sequence. In view of all these factors it is then unsurprising that there is wide variation in the number and importance of direct readout interactions that have been observed in different HTH complexes. The major role that can be occasionally played by indirect readout is strikingly demonstrated in the structure of the *trp* repressor/operator complex.[7] This complex has a large number of contacts between side chains and the phosphate groups of the operator DNA, yet very surprisingly has no functionally significant direct readout base–amino acid interactions at all. Despite this, the cumulative effect of the indirect readout contacts made by the *trp* repressor is to effectively read the particular structural details of its DNA operator sequence.[13]

Readout, both direct and indirect, ensures that key residues on both protein and DNA, are effectively recognized. The functional importance of such residues and hence the protein–DNA contacts involved, can be evaluated by mutagenesis experiments, as well as by surveying their occurrence within a particular family of proteins and their consensus operator sequences. For example, residues tryptophan-48, phenylalanine-49, asparagine-51, and arginine-53 occur in every member of the large eukaryotic homeodomain family. The crystal structure of the engrailed homeodomain–DNA complex[12] shows that the first two play critical roles in preserving the hydrophobic core of the recognition helix, and the other two directly interact with base and phosphate, respectively.

7.3 Major-groove interactions—the α-helix as the recognition element

7.3.1 The helix-turn-helix motif

A protein α-helix can fit snugly only into the major groove rather than the minor groove of a B-like DNA duplex and, therefore, act directly as the recognition element. This element was first observed in the bacterial *cro* repressor protein[14] as part of the helix-turn-helix (HTH) motif. This HTH domain pattern has subsequently been found in the crystal and solution NMR structures of a large number of prokaryotic and eukaryotic transcription regulatory and related proteins, and by comparative sequence analysis in many others. The HTH motif consists of ~20 amino acid residues with residues 1–7 forming the first α-helix and 12–20 the second, linked by a short turn so that the two helices are inclined at 120° to each other. The second helix is the recognition one, which tends to make most of the specific contacts to DNA and lies in the major groove. There are characteristic hydrophobic residues at positions 4, 8, 10, 16, and 18 of the recognition helix, which help to form the overall hydrophobic core of the protein. The N-terminal region of homeodomains typically binds in the DNA minor groove. The resulting side chain interactions with minor groove base edges are important contributors to overall homeodomain binding, with the basic side chain of, in particular, arginine closely mimicking the behaviour of basic side chains in small molecules complexed to the DNA minor groove.

The HTH motif held together in a variety of ways. Thus, the eukaryotic homeodomain engrailed[12,15] has a three-helix bundle (Fig. 7.6), with the HTH motif itself occurring

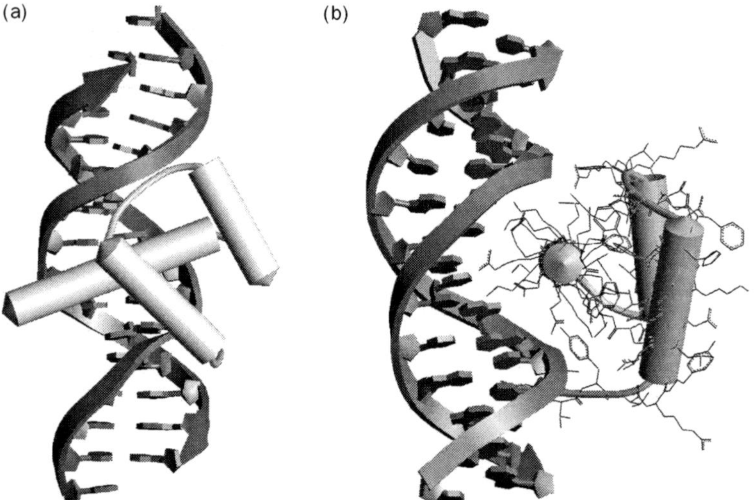

Fig. 7.6 (a) The HTH recognition of the bases in the DNA major groove by the engrailed homeodomain;[12,15] (b) a second view, now looking down the recognition helix of the homeodomain, and showing the side chains extending into the major groove.

Table 7.2 Selected HTH structures

Protein	NDB ID
434 repressor	PDR001
λ repressor	PDR003
Lac repressor	PD0118
Cro repressor	PD0022
Engrailed homeodomain	PD0016
Antennapedia homeodomain	PDR055
MATa1/MATα homeodomain	PDR049
CAP	PDR006

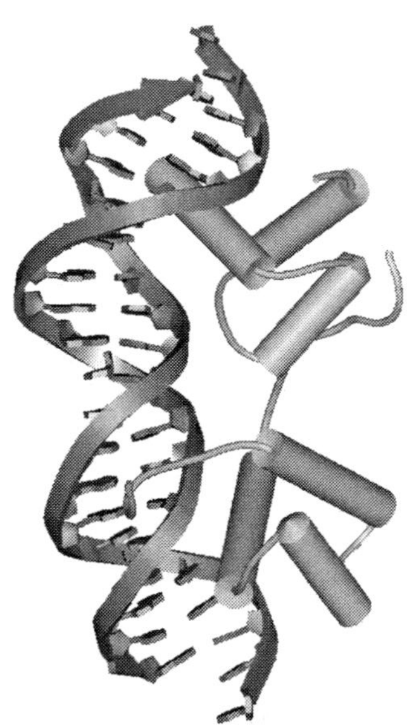

Fig. 7.7 A view of the structure of the DNA–434 repressor complex.[16] The recognition helices are shown oriented into the major groove.

after the first helix, whereas the phage λ[8] and 434 repressor[16] structures are five-helix bundles with the HTH motif comprising helices 2 and 3 (Fig. 7.7). An important difference between the bacterial and eukaryotic HTH proteins is that the former generally bind to their target sites as dimers whereas the latter bind as monomers. This difference is reflected in the structures of the proteins themselves, with some aspects of the non-HTH domains being responsible for dimerization. These structures as seen in the crystalline state, are largely preserved in solution, although side chain orientations are unsurprisingly, not always identical (Table 7.2).[17,18]

Many homeodomains act at a DNA locus together with a second homeodomain, which can be binding in tandem at a neighbouring site. This has been studied in detail for the MATα1 and MATα2 proteins, which regulate transcription in yeast. The crystal structure of the DNA complex with MATα2 alone[15] shows a typical homeodomain, with a straight B-DNA helix. By contrast, the ternary complex[19] has the two protein units contacting each other through the C-terminus tail of the MATα2 domain. This occurs as a consequence of the DNA bending by 60°. A subsequent analogous crystal structure,[20] this time using an A tract DNA, shows a very similar degree of bending, with the A tract bent in the minor-groove direction. Remarkably, the minor-groove spine of hydration is preserved in both structures. Similar bending has been observed with a ternary complex of MATα2, the transcription factor MCM1 and a 26-base pair DNA sequence.[21] Other types of ternary complex can involve the two domains operating in close tandem to overlapping DNA sites, such as in the HoxB1-Pbx1[22] and Ubx-Exd[23] heterodimer structures. Each protein in these complexes induces a small (10°–11° in the former case) bend in the bound DNA, but since the binding sites are close and almost opposite each other, the net effect is an almost straight B-like helix. A general principle emerges from these studies, that when two protein domains bind on the same side of a DNA sequence, then bending ensues; when they are opposite, then there is no net bending.

The majority of HTH proteins have a number of direct interactions between the recognition helix and the major groove of the operator sequence DNA, which maintains a B-type conformation. However, our knowledge of the detailed nature and extent of these interactions is critically dependent on reliable crystal and NMR structures. This caveat is illustrated by the 2.2 Å re-determination[24] of the original 2.8 Å engrailed homeodomain crystal structure.[12] The more recent analysis confirms all major features of the complex, and provides unequivocal information on the role of the important residue Gln50, which previously had been shown to interact only indirectly with a major-groove base edge. This role is confirmed in the higher resolution structure, which also shows several water-mediated contacts between this residue and three (T4, T5, and G7) out of the seven bases in the d(TAATTAC) recognition site. The analysis also shows that there are important contacts between homeodomain side chains and the phosphate backbone, some of which are mediated by water molecules. The case for the key role of water molecules in homeo-domain recognition[25] has been reinforced by molecular dynamics simulations[26] and chemical modification studies.[27] The simulations show that water molecules can have an appreciable residence time at the protein–DNA interface, sufficient for them to be mediating between DNA and amino acid side chains.

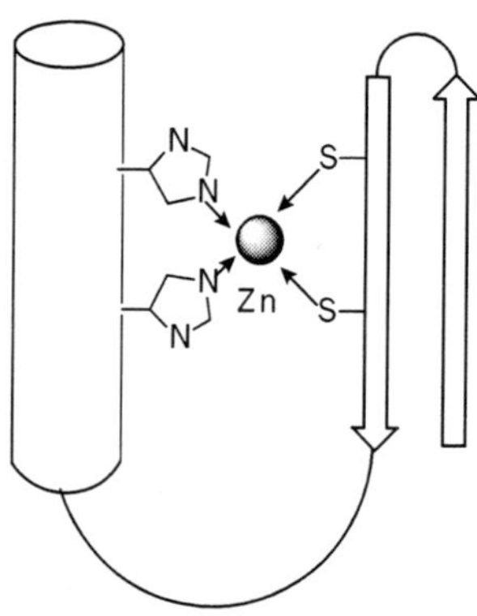

Fig. 7.8 The zinc finger motif, showing the arrangement for a single finger, with the coordination to the zinc atom from histidine residues attached to the α-helix on the left, and the cysteine residues from the β-sheet on the right.

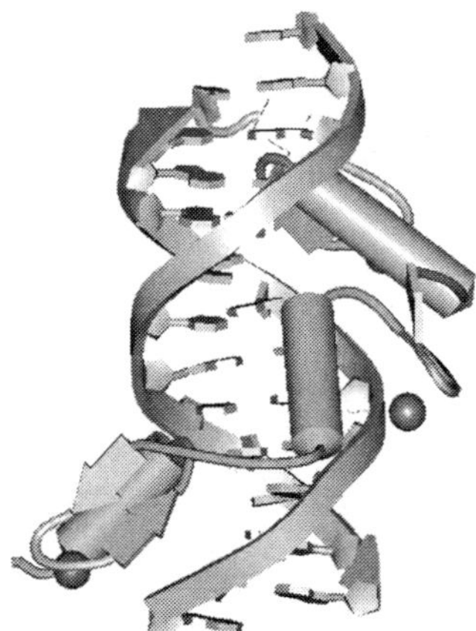

Fig. 7.9 The structure of the Zif268 transcription factor–DNA complex.[31] The zinc atoms in the three zinc fingers are shown as spheres.

7.4 **Zinc-finger recognition modes**

Several zinc-containing motifs for sequence-specific DNA recognition are known, and their details have been revealed by crystallographic and NMR studies. They constitute the largest family of DNA-binding motifs, and show considerable diversity in their architecture. A large group (and the first to be described in the *Xenopus* transcription factor TFIIA) in this family is of the zinc-finger proteins that contain units of regularly spaced cysteine and histidine residues coordinated to a zinc ion.[28–30] Each finger consists of an anti-parallel β sheet and an α helix, held together by the zinc ion coordination (Fig. 7.8).[24] Zinc-finger proteins have at least two such units ('fingers'), with each recognizing a three-base pair site in the major groove of a B-DNA helix, as observed in the crystal structure of the Zif268 transcription factor[31] (Fig. 7.9 and Table 7.3). Each finger makes direct contacts to the guanine-rich strand, with patterns of side chain interactions from the α-helix involving arginine · · · guanine and histidine · · · guanine recognition that are very similar to those seen in HTH proteins.

Table 7.3 Selected zinc-finger protein–DNA structures

Protein	NDB ID
Zif268	PDT006
GAL4	PDT003
Estrogen receptor	PDRC03
Retinoic acid receptor	PDR021
TATA box-zinc finger	PD0186

A quite distinct zinc domain is involved in the structure of the yeast transcription activator GAL4,[32] which bind as a dimer. Each individual GAL4 molecule has two
domains, a zinc-binding region and an α-helix, linked by an extended length of peptide
(Fig. 7.10). The zinc-containing domain rests in the DNA major groove, where there is a
pattern of direct interactions between basic lysine side chains and guanine/cytosine bases
that exactly specifies the highly conserved sequence CCG at this point along the DNA.
Nuclear receptors, such as those for steroid hormones and retinoic acid, also use zinc-
containing motifs for recognition of their DNA response elements, and several NMR and
crystal structures of such complexes have been reported. Almost all the response elements
contain the half-site consensus sequences d(AGGTCA) or d(AGAACA), with the receptors binding as homo- or heterodimers. The zinc motif in them consists of a pair of α-helices
linked by zinc coordination through C-terminal loops (Fig. 7.11). One helix binds in the
DNA major groove at each half-site (Fig. 7.12), while the role of the second helix is to maintain the overall structure.[33–36] The side chains from the recognition helix participate in
normal direct readout interactions; the DNA itself retains B-DNA form, though typically
local distortions to roll and propeller twist have been observed.

Zinc fingers are being used in the design of artificial DNA recognition molecules (proteins?) with sequence specificity that can be altered at will.[37] This approach complements
that of recognition of the minor groove by dimeric polyamides (see Chapter 5), and has the
potential advantage of greater synthetic accessibility, and with applicability to all possible
target sites with equal facility. A peptide containing three zinc fingers, found by screening a
library of peptides, has been shown to bind to a nine-base pair sequence in the BCR–ABL
oncogene,[38] with an equilibrium constant of 6×10^{-7} M. This binding affinity is at least an

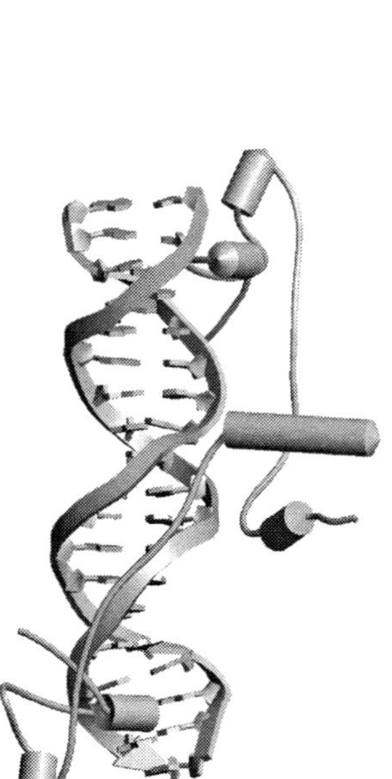

Fig. 7.10 The structure of the
GAL4 transcription factor–DNA
complex.[32]

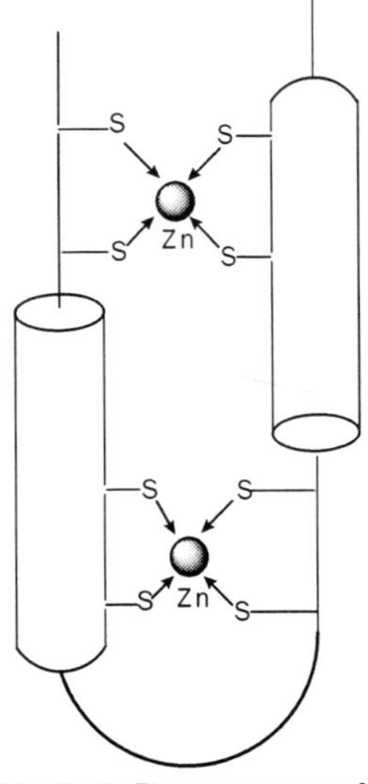

Fig. 7.11 The arrangement for
two zinc fingers as found in the
nuclear receptor family of
transcription factors. The fingers
are formed between helices and
loops, and the zinc atoms are
coordinated by cysteine residues.

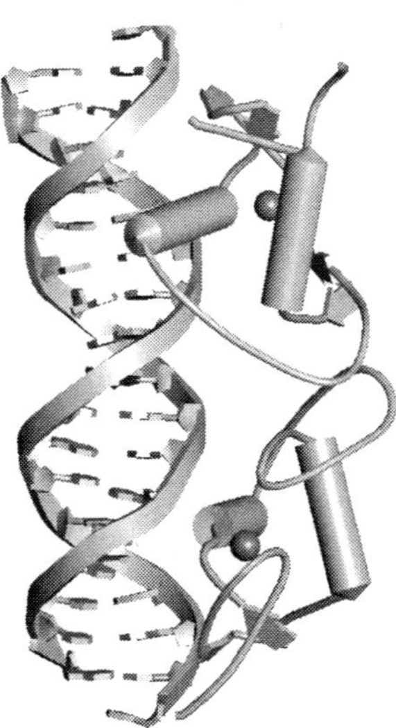

Fig. 7.12 The structure of the
estrogen receptor–DNA complex,[34]
with the zinc atoms in the two zinc
fingers shown as shaded spheres.

order of magnitude higher than to other control sequences. The original strategy did not effectively recognize all possible bases within a triplet, with the 5′ one being optimally adenine or cytosine, possibly as a result of sequence context effects. This problem has now been overcome[39,40] by detailed considerations of several zinc-finger crystal structures that show synergistic recognition from adjacent zinc fingers. As a result, it is possible to start devising a general recognition code for zinc-finger directed DNA sequence specificity. Addition of a dimerization motif can enable more than three zinc fingers to be assembled together without loss of affinity. For example, a four-finger protein has been successfully constructed[41] using the GCN4 leucine zipper dimerization motif (see below), which recognizes 10 base pairs. The challenge for the future is to be able to specifically recognize much longer DNA sequences by this strategy.

7.5 Other major-groove recognition motifs

A number of eukaryotic transcription factors contain the basic 'leucine-zipper' recognition and dimerization element, as found in the yeast transcription factor GCN4,[42] which consists of a very long continuous, almost straight α helix with regularly repeating leucine residues (Fig. 7.13). Two such helices interact together to form a parallel coiled coil. There are a large number of interactions with the DNA backbone from the basic side chains, as well as specific, direct readout ones. The orientation of the α-helix within the DNA major groove is critical for these contacts to take place, with the helix being approximately parallel to the phosphodiester backbone (Table 7.4).

A distinct DNA-binding motif is used by the transcription activator E2 from papilloma virus,[43] with a dimeric antiparallel β-barrel which delivers a pair of α-helices to the major groove. Protein–DNA contacts involve direct side chain interactions with the bases, as well as indirect backbone contacts. A distinguishing feature of the structure is the smooth bending of the DNA around the β-barrel, with a radius of curvature of 45 Å. This is less extreme than the 90° bend found[44] in the DNA of the complex with the dimeric HTH domain *E. coli* transcription activator protein (CAP). The bending in the CAP complex, discussed later, is due to a large number of interactions with DNA

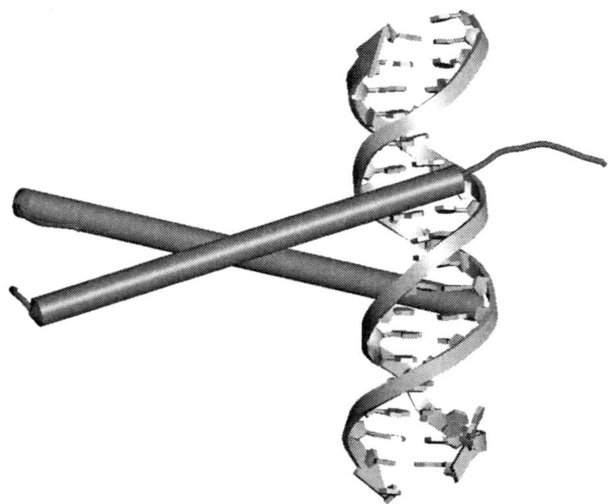

Fig. 7.13 The structure of the yeast transcription factor GCN4–DNA complex.[42]

Table 7.4 Structures of protein–DNA complexes
with miscellaneous binding motifs

Protein	NDB ID
Met J	PD0245
E2	PDV001
P53	PDR022
Ku	PD0220
NF-kB	PDR051
Nucleosome	PD0001

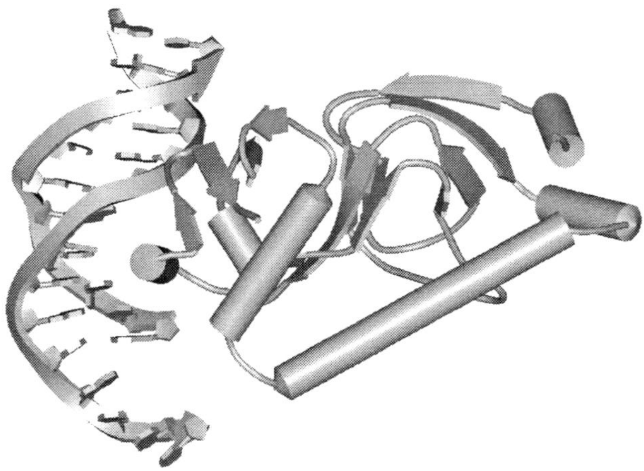

Fig. 7.14 The structure of the
E. coli transcription activator
protein (CAP)–DNA complex,
showing one CAP unit with its
recognition helix in the major
groove of the bent DNA.[44]

phosphate groups, which serve to position the recognition helix correctly in the major groove (Fig. 7.14). A totally distinct non-helix recognition motif has been found in the structures of a few bacterial repressors, typified by the *met* J repressor/operator complex,[45] where double-stranded anti-parallel β-ribbons are the major-groove recognition elements. Side chains from the ribbons interact directly with the DNA phosphates and bases, the side chain · · · base direct readout mechanism remains the same as that observed with the other repressor/operator complexes outlined above. Other analogous types of folds have been found in the structures of several DNA repair enzymes.

7.6 **Minor-groove recognition**

7.6.1 **Recognition of B-DNA**

Sequence-specific proteins generally do exploit the major groove to a greater extent than the minor one. This is not always solely on account of its greater information potential. The ability of an α-helix to snugly fit into the major groove is also a factor. This led, in the past, to a view that the minor groove is of little importance. However, there are now a large number of examples where interactions of side chains in the minor groove are significant components of the overall protein–DNA stabilization (Table 7.5). An arginine

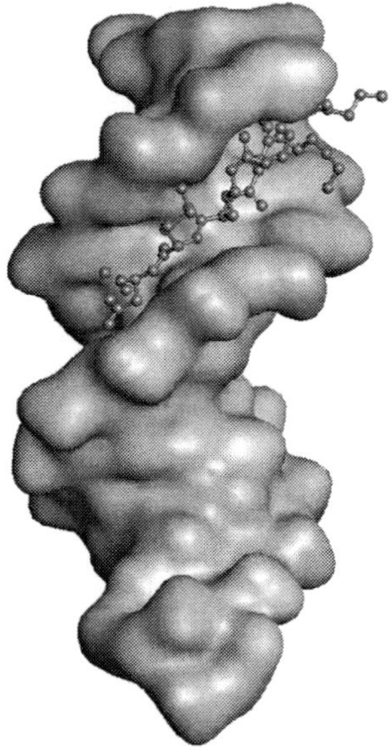

Fig. 7.15 A view of the structure of the Hin recombinase–DNA complex, showing the N-terminus residing in the minor-groove surface of the DNA.[47]

Table 7.5 Structures of selected protein–DNA complexes involving minor-groove recognition

Protein	NDB ID
Hin recombinase	PDE009
TBP	PD0070, PD0154
TBP–TFIID complex	NBP0228
TBP–TFIIB complex	PDT032
HMG-D	PD0110

side chain of the 434 repressor[16] bridges to bases and phosphate groups via water molecules. Homeodomain structures[12,15,19–21] have extended N-terminal arms which lie in the DNA minor groove, making extensive base and backbone contacts to the A/T regions of the operator sequence. The hydrogen-bonding interactions that the arginine side chains in these complexes make with the O2 atom of thymines, are closely analogous to the mode of interaction shown by the minor-groove binding drugs. These together with the indirect readout of the dimensions of the groove itself by van der Waals interactions involving the side chains, is a significant factor in the preference shown by homeodomains for A/T-rich sites on DNA, again analogous to the preferences shown by the drugs. These N-terminal sequences are also related to the proposal[46] that the repeating sequence SPKK (serine–proline–lysine/ arginine–lysine/arginine), which occurs in, for example, the N-terminus of some histone proteins, also binds in the minor groove at A/T-rich regions.

The crystal structure of the HTH-motif Hin recombinase enzyme bound to a 14 base pair A/T-rich DNA sequence[47] shows that the straight B-DNA helix has a narrow minor groove in the A/T region, in which the N-terminus of the enzyme resides (Fig. 7.15). There are several amino acid–base edge specific contacts, such as an arginine to N3 of an adenine. More unusually, the main chain amide of this arginine is in hydrogen-bonding contact with O2 of a thymine. The pattern of hydrogen bonding in this A/T-rich region is reminiscent of netropsin binding. The C-terminus of this enzyme also lies in the minor groove, entering at the other end of the DNA duplex.

Loops can also bind in the minor groove, as found in the solution (NMR) structure of the Mu repressor protein–DNA complex,[48] which has an HTH motif containing an additional 'wing' loop between helix 2 and 3. The wing is flexible in the absence of DNA, but in the complex is found inserted in the minor groove, where two lysine side chains make extensive contacts with, in particular, adenine and thymine base edge atoms.

7.6.2 **The opening-up of the minor groove by TBP**

The general transcription factor complex TFIID plays a key role in the initiation of transcription in eukaryotic cells. It functions by binding a component protein, TBP, to the 'TATA box' sequence upstream of the start of transcription. This has been shown by chemical protection studies to bind in the minor groove at the 5'-TATA site itself, as well as in the major groove on the 3′ side of this site. There is a strong preference for 5′-TATA as compared to other A/T-containing sequences. The crystal structure of the TBP protein,[49] in the absence of DNA, shows a novel DNA-binding fold, with a symmetric α/β arrangement. It was initially suggested that that the saddle-shaped arrangement of the α/β structure, with an extended concave surface, is ideally complementary to a B-form DNA duplex, and would be effectively wrapped around it. Mutagenesis studies have identified the key protein residues involved in DNA binding. They are arranged on the concave surface of the protein, as suggested by this interaction model.

However the actual structures of the TBP–DNA complexes shows a very different arrangement for the DNA compared to this initial model.[50,51] The DNA is dramatically bent, by c. 80° so as to follow the concave curvature of the β-sheet, i.e., it is at right angles to the earlier model (Fig. 7.16). This structure has been observed in TATA-containing DNA sequences complexed with TBP from a wide range of organisms, so is clearly a general feature of TATA box recognition.[52] Eight base pairs are in contact with the β-sheet saddle, and are in a non-B-DNA conformation. By contrast, analysis of complexes with longer DNA sequences shows that the DNA flanking the central site is in a B-like form, on both 3′ and 5′ sides. Structure determination at high-resolution (1.9 Å) of a TBP complex with the sequence d(TATAAAAG) has enabled a detailed view of the bent DNA structure to be obtained.[53] The DNA is unwound by 105° over the seven base pairs, and has a greatly enlarged and flattened minor groove (with a maximum width of over 9 Å) to accommodate the β-sheet saddle. The major groove is highly compressed. There are large positive rolls, of up to 40° at the TA step in the 5′-TATA sequence, together with large propeller twists, of up to −39° in the 5′-AAAA sequence. These large twists result in a pattern of A tract major-groove bifurcated hydrogen bonds. Sugar puckers are in the C3′-*endo* family. The overall impression of the DNA conformation in the TATA box region is of distorted

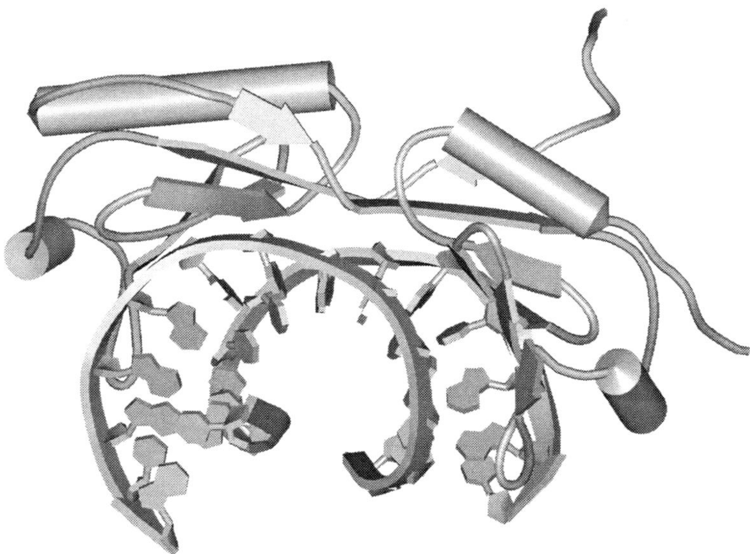

Fig. 7.16 The structure of the TBP–DNA complex.[53] Note the large curvature of the DNA duplex.

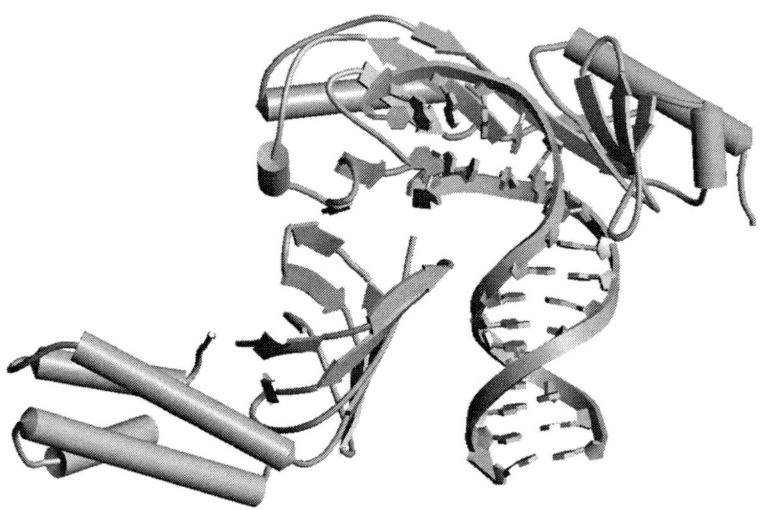

Fig. 7.17 Structure of the TFIIA–TBP–DNA complex.[55] Note that the DNA reverts to B-form outside the TBP saddle structure.

A-type form, with abrupt changes to B-DNA immediately outside the box. The bending is achieved by pairs of phenylalanine residues that are inserted into the ends of the TATA box, and are stacked with several bases and base pairs. These bases become unstacked from the DNA helix and result in the observed kinking. There are few direct readout base–side chain contacts, but numerous amino acid contacts with phosphate groups. Many of these, especially with lysine side chains, are water-mediated.

Several ternary complexes of general transcription factors with the TBP–TATA complex have been reported. Those with TFIIA[54] and TFIIB[55] show that the TBP–TATA box structure seen in the binary complexes is fully retained, strongly suggesting that these transcription factors bind to a preformed TATA-box complex (Fig. 7.17), as does the negative cofactor NC2 in its TBP–DNA complex.[56]

7.6.3 **Other proteins that induce bending of DNA**

The ability of TBP to bend DNA on binding to its recognition sequence is shared by a number of other minor-groove regulatory proteins notably those containing the so-called HMG (high mobility group) box sequence-neutral DNA-binding domains. Such proteins include SRY, which determines the expression of human male-specific genes. NMR studies of HMG boxes [57–59] show that the structure consists of three α-helices arranged in an L shape, with conserved basic arginines and lysines on the inner face of the L. The helices do not correspond to those in any (major-groove binding) HTH arrangement, and thus constitute a novel DNA-binding motif. A crystal structure of a complex between HMG-D and an A/T-rich duplex decamer[60] shows (Fig. 7.18) that the bound DNA is very distorted from canonical B-DNA, and smoothly bent. The protein binds to the minor groove, which is underwound, widened and flattened so that the overall shape resembles A-DNA. This is reminiscent of the distortions induced by the TBP protein, and indeed there are significant similarities (Fig. 7.19), although here the minor groove is even wider, 12.4 Å.

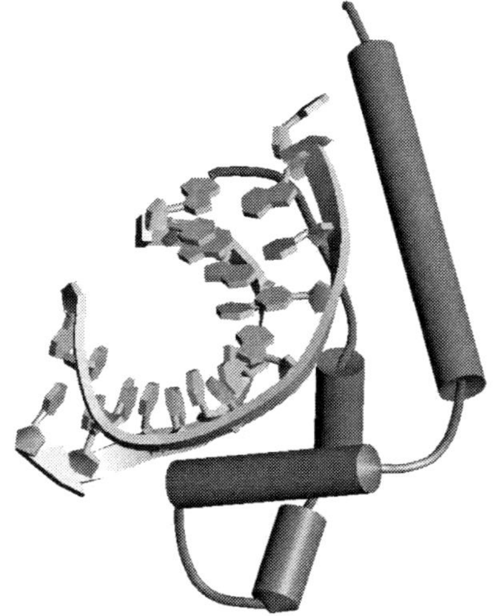

Fig. 7.18 Structure of the
HMG–D-DNA complex.[60]

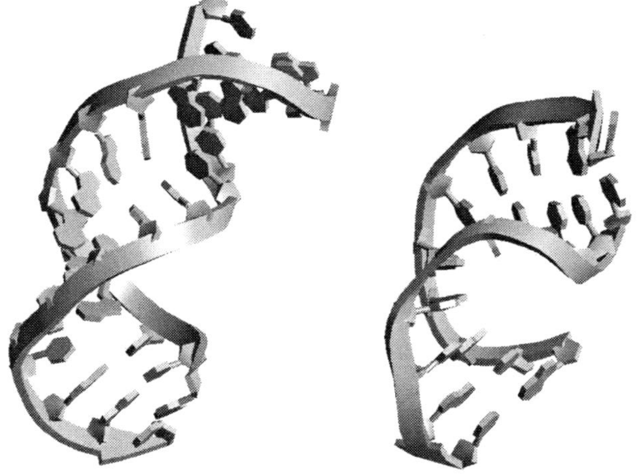

Fig. 7.19 Comparative views of
the bent DNA in the TBP–TATA
complex[53] and in the
HMG–D-DNA complex.[60]

The organization of almost two metres of DNA in the eukaryotic chromosome requires that it is packaged up in a very tight manner.[61] The basic unit of packaging is the nucleosome core particle, which consists of an octamer of pairs of the four histone proteins H2A, H2B, H3 and H4, together with, between 160 and 240 DNA base pairs, wound around the core. The nucleosome particle itself is the repeating unit for the further level of supercoiled organization in chromosomes. The crystal structure of the nucleosome core particle was initially determined at 7 Å resolution,[62] and more recently at 2.8 Å,[63] which has enabled details such as DNA bending, to be described. In this structure, there are 146 base pairs, which are wound around the octameric histone core, forming a flattened superhelix (Fig. 7.20). There are 10.2 base pairs per turn and the bending is altogether regular, with the DNA retaining B-like features. The smooth bending is accomplished by relatively small but numerous deformations from ideality, in a sequence-dependent manner, similar to changes observed in crystal structures of some short, yet bent oligonucleotides.

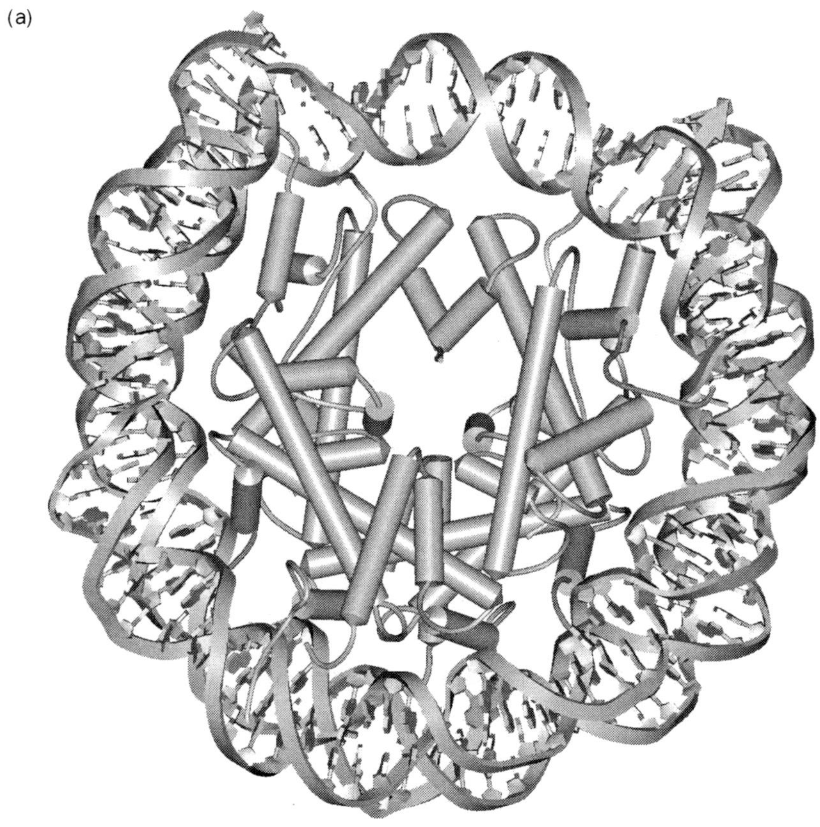

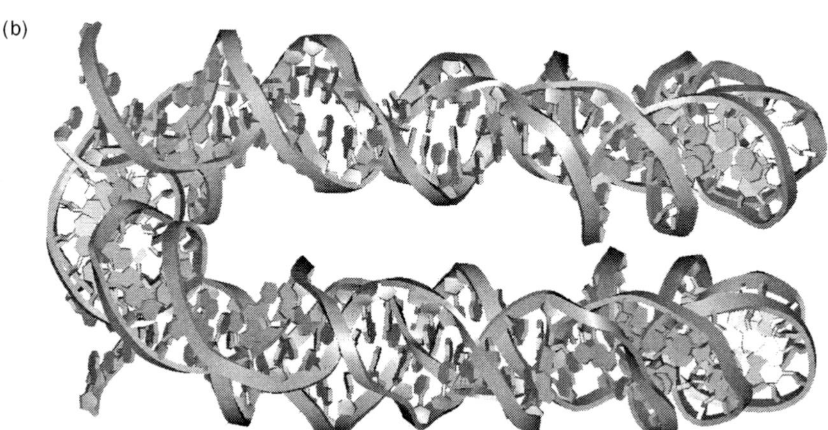

Fig. 7.20 Two views of the nucleosome structure,[63] showing (a) the histone octamer surrounded by supercoiled DNA; (b) shows a view at right angles highlighting the disc-like nature of the almost complete two turns of bound supercoiled DNA.

7.7 **DNA bending and protein recognition**

X-ray crystallography and other techniques have shown that DNA structures in the absence of bound drug or protein can have a significant degree of flexibility. This is reflected in sequence-dependent backbone, sugar pucker, base step and base pair features, although the classic features of canonical forms are often observed in both native and protein-bound oligonucleotide structures.

There is considerably greater diversity of DNA structure in protein complexes than in native structures, largely because the binding energy for a protein–DNA interaction is sufficient to readily overcome the barriers to local DNA deformability. Hence these can lead to even large-scale changes in DNA structure such as bending, so as to optimize interactions, should such changes be needed. Many proteins, notably HTH and zinc fingers, do not need to distort in order to produce the required protein contacts. Sometimes though when distortions do occur they can blur the distinctions between B- and A-form DNA structural types so that it is no longer possible to unequivocally assign the resulting DNA structure to one or the other form. This is especially apparent when large changes in parameters such as groove width, helical twist and base pair inclination to the helix axis are observed. For example, the nature of the structure of DNA in the Zif268–DNA zinc finger complex has been controversial, with both A- and B-type helices being suggested. The structure of the complex[31] shows that the DNA adopts an overall B-type form, although there are a number of local features that are intermediate between A and B. In such a case, the simple canonical DNA classifications are of limited use by themselves and are best accompanied by more detailed information describing particular features.

The DNA in the *met* J complex[45] has a straight 10 base pair central segment, with bends at each end as a result of groove width changes necessary for both effective protein dimer formation and protein–DNA interaction. In the structure of the *trp* repressor/operator complex[7,13] the DNA is appreciably distorted from canonical B-form as a result of numerous small, cumulative local changes to, in particular, roll and slide. These together with changes in groove width and backbone geometry from canonical B-DNA are necessary in order to achieve the large number of indirect readout contacts that produce specificity. The DNA is bent and its surface closely follows the contours of the protein surface. Rather greater DNA bending is required in the papillomavirus-1 E2 DNA complex,[43] where there is smooth DNA bending over the β barrel recognition motif, in order to expose the major groove to protein side chains. Groove widths are compressed on the side of the DNA that faces protein, with alterations in propeller twist and roll angles contributing to these width changes. The prokaryotic transcription activator CAP, which acts as a dimer, has a consensus DNA binding site 22 base pairs long with the essential bases towards the end of the sequence. This is far too long a DNA sequence to be contacted by the relatively small CAP dimer on the basis of the DNA remaining linear. The structure of the CAP–DNA complex shows[44] that the DNA has solved this problem by bending by ~90° in order to effectively wrap around the protein dimer. This bending, or kinking, is produced almost entirely at two base pair steps, one on each side of the 2-fold axis of the complex, by roll angles of ~40°. It is likely that this bending is an important aspect of transcriptional activation by CAP and other analogous proteins. There is a very close structural correspondence between CAP and other architectural proteins[64] such as TBP and the globular domain of histone H5[65] (which is involved in higher level nucleosome organization). These suggest that the ability of this particular type of HTH protein to bend DNA sequences is of wide structural and functional significance.

In the light of the observations of DNA deformability in these and other protein complexes, the question arises as to their relevance to the numerous studies on native DNA structures. The answer is not clear-cut, and depends on the particular structures involved. It is apparent though that native structures do sometimes demonstrate inherent sequence-dependency which is also found in their protein complexes, provided the changes in structure are not profound. An example is provided by the *trp* operator site,[66] which shares common features of structure and hydration in the native structure and in the protein complex. Larger scale DNA conformational differences such as are seen in the TBP and other minor-groove binding protein complexes, can still be related to native DNA structures since these are often describable in terms of simple concerted changes from standard forms of DNA. Thus, the structure of the DNA in the TATA box may be thought of as canonical A form with just a change in glycosidic angle.[67] It is striking that this and other small variants on A form,[68] when embedded within B-DNA, can produce bent and curved structures suitable for binding to a wide range of proteins (and small molecules such as *cis*-platinum—see Chapter 5). Bending was foreseen over 20 years ago, as a natural consequence of junctions between canonical A and B forms.[69] It was less obvious until comparatively recently, that *both* forms would play key roles in the functioning of DNA.

References

1 See the special journal issues devoted to the announcement of the sequence of the human genome: International Human Genome Sequencing Consortium, (2001) *Nature*, **409**, 860; *Science*, **291**, 1304.

2 Seeman, N. C., Rosenberg, J. M., and Rich, A. (1976). *Proceedings of the National Academy of Sciences USA*, **73**, 804.

3 Luscombe, N. M., Austin, S. E., Berman, H. M., and Thornton, J. M. (2000). *Genomic Biology*, **1**, 1.

4 Harrison, S. C. (1991). *Nature*, **353**, 715.

5 Nadassy, K., Wodak, S. J., and Janin, J. (1999). *Biochemistry*, **38**, 1999.

6 Luscombe, N. M., Laskowski, R. A., and Thornton, J. M. (2001). *Nucleic Acids Research*, **29**, 2860.

7 Otwinowski, Z., Schevitz, R. W., Zhang, R.-G., Lawson, C. L., Joachimiak, A., Marmorstein, R. Q., *et al.* (1988). *Nature*, **335**, 321.

8 Beamer, L. J. and Pabo, C. O. (1992). *Journal of Molecular Biology*, **227**, 177.

9 Matthews, B. W. (1988). *Nature*, **335**, 408.

10 Mandel-Gutfreund, Y., Schueler, O., and Margalit, H. (1995). *Journal of Molecular Biology*, **253**, 370.

11 Pabo, C. O. and Nekludova, L. (2000). *Journal of Molecular Biology*, **301**, 597.

12 Kissinger, C. R., Liu, B., Martin-Blanco, E., Kornberg, T. B., and Pabo, C. O. (1990). *Cell*, **63**, 579.

13 Haran, T. E., Joachimiak, A., and Sigler, P. B. (1992). *The EMBO Journal*, **11**, 3021.

14 Anderson, W. F., Ohlendorf, D. H., Takeda, Y., and Matthews, B. W. (1981). *Nature*, **290**, 744.

15 Wolberger, C., Vershon, A. K., Liu, B., Johnson, A. D., and Pabo, C. O. (1991). *Cell*, **67**, 517.

16 Aggarwal, A. K., Rodgers, D. W., Drottar, M., Ptashne, M., and Harrison, S. C. (1988). *Science*, **242**, 899.

17 Neri, D., Billeter, M., and Wüthrich, K. (1992). *Journal of Molecular Biology*, **223**, 743.

18 Fraenkel, E. and Pabo, C. O. (1998). *Nature Structural Biology*, **5**, 692.

19 Li, T., Stark, M. R., Johnson, A. D., and Wolberger, C. (1995). *Science*, **270**, 262.

20 Li, T., Jin, Y., Vershon, A. K., and Wolberger, C. (1998). *Nucleic Acids Research*, **26**, 5707.

21   Tan, S. and Richmond, T. J. (1998). *Nature*, **391**, 660.

22   Piper, D. E., Batchelor, A. H., Chang, C.-P., Cleary, M. L., and Wolberger, C. (1999). *Cell*, **96**, 587.

23   Passner, J. M., Ryoo, H. D., Shen, L., Mann, R. S., and Aggarwal, A. K. (1999). *Nature*, **397**, 714.

24   Fraenkel, E., Rould, M. A., Chambers, K. A., and Pabo, C. O. (1998). *Journal of Molecular Biology*, **284**, 351.

25   Gehring, W. J., Qian, Y. Q., Billeter, M., Furukubo-Tokunaga, K., Schier, A., Resendez-Perez, D., *et al.* (1994). *Cell*, **78**, 211.

26   Billeter, M., Güntert, P., Luginbühl, P., and Wüthrich, K. (1996). *Cell*, **85**, 1057.

27   Labeots, L. A. and Weiss, M. A. (1997). *Journal of Molecular Biology*, **269**, 113.

28   Klug, A. and Rhodes, D. (1987). *Trends in Biochemical Science*, **12**, 464.

29   Laity, J. H., Lee, B. M., and Wright, P. E. (2001). *Current Opinion in Structural Biology*, **11**, 39.

30   Wolfe, S. A., Nekludova, L., and Pabo, C. O. (2000). *Annual Reviews in Biophysics*, **29**, 183.

31   Pavletich, N. P. and Pabo, C. O. (1991). *Science*, **252**, 809.

32   Marmorstein, R., Carey, M., Ptashne, M., and Harrison, S. C. (1992). *Nature*, **356**, 408.

33   Luisi, B., Xu, W. X., Otwinowski, Z., Freedman, L. P., Yamamoto, K. R., and Sigler, P. B. (1991). *Nature*, **352**, 497.

34   Schwabe, J. W. R., Chapman, L., Finch, J. T., and Rhodes, D. (1993). *Cell*, **75**, 567.

35   Rastinejad, F., Perlmann, T., Evans, R. M., and Sigler, P. B. (1995). *Nature*, **375**, 203.

36   Zhao, Q., Chasse, S. A., Devarakonda, S., Sierk, M. L., Ahvazi, B., and Rastinejad, F. (2000). *Journal of Molecular Biology*, **296**, 509.

37   Choo, Y. and Isalan, M. (2000). *Current Opinion in Structural Biology*, **10**, 411.

38   Choo, Y., Sánchez-García, I., and Klug, A. (1994). *Nature*, **372**, 642.

39   Isalan, M., Choo, Y., and Klug, A. (1998). *Biochemistry*, **37**, 12026.

40   Isalan, M., Klug, A., and Choo, Y. (2001). *Nature Biotechnology*, **19**, 656.

41   Wolfe, S. A., Ramme, E. I., and Pabo, C. O. (2000). *Structure*, **8**, 739.

42   Ellenberger, T. E., Brandl, C. J., Struhl, K., and Harrison, S. C. (1992). *Cell*, **71**, 1223.

43   Hegde, R. S., Grossman, S. R., Laimonis, L. A., and Sigler, P. B. (1992). *Nature*, **359**, 505.

44   Schultz, S. C., Shields, G. C., and Steitz, T. A. (1991). *Science*, **253**, 1001.

45   Somers, W. S., and Phillips, S. E. V. (1992). *Nature*, **359**, 387.

46   Suzuki, M. (1988). *The EMBO Journal*, **8**, 797.

47   Feng, J. A., Johnson, R. C., and Dickerson, R. E. (1994). *Science*, **263**, 348.

48   Wojciak, J. M., Iwahara, J., and Clubb, R. T. (2001). *Nature Structural Biology*, **8**, 84.

49   Nikolov, D. B., Hu, S.-H., Lin, J., Gasch, A., Hoffmann, A., Horikoshi, M., *et al.* (1992). *Nature*, **360**, 40.

50   Kim, J. L., Nikolov, D. B., and Burley, S. K. (1993). *Nature*, **365**, 520.

51   Kim, Y., Geiger, J. H., Hahn, S., and Sigler, P. B. (1993). *Nature*, **365**, 512.

52   Patikoglou, G. A., Kim, J. L., Sun, L., Yang, S.-H., Kodadek, T., and Burley, S. K. (1999). *Genes and Development*, **13**, 3217.

53   Kim, J. L. and Burley, S. K. (1994). *Nature Structural Biology*, **1**, 638.

54   Nikolov, D. B., Chen, H., Halay, E. D., Usheva, A. A., Hisatake, K., Lee, D. K., *et al.* (1995). *Nature*, **377**, 119.

55   Tan, S., Hunziker, Y., Sargent, D. F., and Richmond, T. J. (1996). *Nature*, **381**, 127.

56   Kamada, K., Shu, F., Chen, H., Malik, S., Steizer, G., Roeder, R. G., *et al.* (2001). *Cell*, **106**, 71.

57   Weir, H. M., Kraulis, P. J., Hill, C. S., Raine, A. R. C., Laue, E. D., and Thomas, J. O. (1993). *The EMBO Journal*, **12**, 1311.

58   Werner, M. H., Huth, J. R., Gronenborn, A. M., and Clore, G. M. (1995). *Cell*, **81**, 705.

59   Allain, F. H.-T., Yen, Y.-M., Masse, J. E., Schultze, P., Dieckmann, T., Johnson, R. C., and Feigon, J. (1999). *The EMBO Journal*, **18**, 2563.

60   Murphy, F. V., Sweet, R. M., and Churchill, M. E. A. (1999). *The EMBO Journal*, **18**, 6610.

61   Luger, K. and Richmond, T. J. (1998). *Current Opinion in Structural Biology*, **8**, 33.

62   Richmond, T. J., Finch, J. T., Rushton, B., Rhodes, D., and Klug, A. (1984). *Nature*, **311**, 532.

63   Luger, K., Mäder, A. W., Richmond, R. K., Sargent, D. F., and Richmond, T. J. (1997). *Nature*, **389**, 251.

64   Werner, M. H. and Burley, S. K. (1997). *Cell*, **88**, 733.

65   Ramakrishnan, V., Finch, J. T., Graziano, V., Lee, P. L., and Sweet, R. M. (1993). *Nature*, **362**, 219.

66   Shakked, Z., Guzikevich-Guerstein, G., Frolow, F., Rabinovich, D., Joachimiak, A., and Sigler, P. B. (1994). *Nature*, **368**, 469.

67   Guzikevich-Guerstein, G. and Shakked, Z. (1996). *Nature Structural Biology*, **3**, 32.

68   Lu, X.-J., Shakked, Z. and Olson, W. K. (2000). *Journal of Molecular Biology*, **300**, 819.

69   Selsing, E., Wells, R. D., Alden, C. J., and Arnott, S. (1979). *Journal of Biological Chemistry*, **254**, 5417.

Further reading

Garvie, C. W. and Wolberger, C. (2001). *Molecular Cell*, **8**, 937.

Harrison, S. C. and Aggarwal, A. K. (1990). *Annual Reviews in Biochemistry*, **59**, 933.

Morávek, Z., Neidle, S., and Schneider, B. (2002). *Nucleic Acids Research*, **30**, 1182.

Pabo, C. O. and Sauer, R. T. (1992). *Annual Reviews in Biochemistry*, **61**, 1053.

Patikoglou, G. and Burley, S. K. (1997). *Annual Reviews in Biophysics*, **26**, 289.

Steitz, T. A. (1990). *Quarterly Reviews in Biophysics*, **23**, 205.

Travers, A. A. (1993). *DNA–protein Interactions*, Chapman and Hall, London.

Useful web sites

http://www.biochem.ucl.ac.uk/bsm/prot_dna/prot_dna.html   This provides summaries of all the protein–DNA families, and the relevant features in many individual structures.

http://www.rtc.riken.go.jp/jouhou/3dinsight/recognition.html   This database provides facilities for a variety of searches, and includes experimental data on binding.

Index